POLE SHIFT

EVIDENCE WILL NOT BE SILENCED

DAVID MONTAIGNE

About the Author

The son of two Philadelphia teachers from different religious backgrounds, David Montaigne was raised to respect ancient wisdom and learned not to gloss over the scientific and mathematical clues in ancient texts. In his everyday life he is a Penn State graduate, father, entrepreneur, a professional electrician, and an amateur cook and fisherman.

Montaigne is also an avid reader, investigator, prophecy scholar, and author of several books. Years ago you may have passed him near the archives in Washington D.C., London, or the Vatican – or shared a bus ride to a pyramid or temple in Israel, Egypt, Turkey, Greece, Mexico or Guatemala… But it is now more likely he is searching for answers online – at home in Lancaster County, Pennsylvania – where he celebrates every possible moment with his three inquisitive children, who remind him that we should never stop asking questions.

Books by this author include:

Nostradamus: World War III

End Times and 2019

Antichrist 2016-2019

Nostradamus And The Islamic Invasion Of Europe

Pole Shift: Evidence Will Not Be Silenced

POLE SHIFT: EVIDENCE WILL NOT BE SILENCED

Copyright © 2018 David Montaigne

All rights reserved.

ISBN: 1986785130

ISBN-13: 978-1986785136

Your government and religious leaders may not want you to know, but the evidence suggests that pole shifts are both magnetic and geophysical, with a periodic cycle of recurring and predictable cataclysms involving huge earthquakes and tsunamis, changes in latitude and altitude, mass extinctions, and the destruction of civilizations, reducing them to myth and legend. Evidence also suggests that the next pole shift is due in the 21st century.

Table of Contents

About the Author - 2

Foreword - 6

Introduction - 8
Mistranslations, Genesis, Uniformitarianism, Catastrophism, Conspiracy, Magnetic Pole Shifts, South Atlantic Anomaly, Atlantis, Antarctica, Mythology, Nibiru

Evidence of Previous Pole Shifts - 34
Coral Equators, Bands of Soil, Lava Rock, Frozen Mammoths, Ice Ages, Mountain Building, Human Records, Ancient Maps, Gothenburg Flip, South America, Usselo Horizon, Waterfalls, Penguins & Seals, Tree Rings, Pole Shifts on Other Worlds

The Great Flood – Was Not Caused by a Pole Shift - 68
Noah, Biblical Chronology, Calendar Beginnings, Taurus & Pleiades, Holidays of Death, Burckle Crater, Mega-tsunami

Ancient Comments on Pole Shifts - 79
Ipuwer, Hesiod, Heraclitus, Herodotus, Plato, Diodorus, Seneca

Early Modern Writers on Pole Shifts - 92
Boehme, Milton, Burnet, Newton, Cuvier, Darwin, Agassiz, Klee, Lubbock, Evans, Drayson, de Bourbourg, Donnelly, Warren, Blavatsky, Wheeler, Churchward, Cayce, Brown, Velikovsky, Hapgood, Einstein

Chan Thomas' Classified Adam and Eve Story - 130
Genesis, CIA Partial Declassification, Speculation, Causes, Timing, Surviving the Pole Shift, Advanced Ancient Civilization, Weaponizing Pole Shifts, Suggestions for America's Enemies, Third Secret of Fatima, Tunguska Blast, Nikola Tesla

Space Age Evidence, Modern Theories, and Ancient Legends - 161
Barbiero, Flem-Ath, Schoch, LaViolette, Carlson, Pole Shift Articles, World Mythology

Pole Shifts in the Bible - 183
Enoch, Elijah, Isaiah, Revelation, Calculating Dates, Samson, Jesus

Nibiru – Fiction and Fact - 216
Sitchin, Planet X, Zetas, Enuma Elish, Sumerians, Babylonians, Astronomy, Crossing Point

World Mythology - 237
Chinese, Greek, Norse, Egyptian, Aztec, Mayan, Indian, Native American, Icelandic, Finnish, Santa, Destroyed Island Homeland, Deity Sacrifice, Unhinged Mill, Avatars of Destruction, Phaethon, Myth and Astronomy, Biblical stories, Ancient Prophecies

Surviving the Pole Shift - 277
Warnings, Timing, Tsunamis, Prepping, Balance

Selected Bibliography - 280

End Notes - 281

Foreword

Pole shifts are global catastrophes in which the surface of the Earth – the entire outer crust of the planet – suddenly moves in one solid piece over the layers of liquid rock below like the chocolate around a cherry. Areas near the North and South Poles move thousands of miles towards the equator, causing ice caps to start melting in warmer latitudes, while equatorial regions along the same line of movement head out of the tropics towards colder climates. Some formerly temperate lands end up at the new poles and suffer a sudden deep freeze, soon to be buried under new polar ice caps.

Lands approaching the equator slip under sea level as they enter the equatorial bulge; other formerly submerged sea beds rise to become dry land for the first time in many thousands of years. Significant changes in latitude and altitude are accompanied by earthquakes and tsunamis of biblical proportions. Countless species of plants and animals go extinct, while many others adapt to the new conditions and thrive in the new world.

This is not some crackpot theory. Books about pole shifts have been written by history professors, electrical engineers and naval admirals. Albert Einstein wrote the foreword to one book describing how mass imbalances "produce a movement of the earth's crust over the rest of the earth's body, and this will displace the polar regions toward the equator."[1] Sir Isaac Newton said such a mass imbalance "by its perpetual endeavour to recede from the centre of its motion will disturb the motion of the globe, and cause its poles to wander about its surface."[2] Dozens of scientists have reported on evidence that pole shifts have happened before and will happen again.

This periodically recurring natural disaster transforms the surface of the Earth and reduces human civilizations to myths and legends. Evidence suggests that pole shifts occur in regular cycles and that we are close to the next pole shift event bringing a sudden end to our present world. Many myths and monuments encode astronomical and numerical clues that may allow us to predict the timing of the next pole shift.

Most people are dimly aware of the evidence of previous pole shifts, but they haven't given the clues much thought. We rarely stop to question anything unexplained for very long – our brains are wired to move along to the next entertaining distraction. Most of us don't treat curiosities as evidence in some scientific mystery that it's our job to solve personally. The world is complicated, and most of us don't feel the need to understand how everything works. But once in a while, someone does make it a personal quest to study seemingly unrelated anomalies as potential puzzle pieces to put together into a theory that answers all the unsolved mysteries. This book is my attempt to prove there is a periodic cycle of regularly recurring pole shifts, bringing together all the evidence of past catastrophes and demonstrating that we are probably due for the next pole shift very soon.

The first person to read my rough draft warned me that many people will not read more than a fraction of my book. Not because it isn't good – but because the evidence I present is so overwhelming that he thinks anyone who reads even fifty pages of material will be convinced that I am right, and that they will not feel the need to read all the evidence. I guess that's good, if my argument is so convincing – but I hope many of you will want to continue through the entire book. If you do, you will see that many questions are solved by the pole shift theory, including:

What causes ice ages? Why do bands of ancient coral crisscross the world like misplaced former equators? Why do we find mammoths that were suddenly frozen with undigested summer vegetation, flowers, and seeds in their mouths and stomachs? Why is the Earth's magnetic field weakening? Why is the magnetic North Pole racing towards Russia? Why do Emperor Penguins march eighty miles inland to lay eggs in the sunless Antarctic winter at forty degrees below zero? Why do ancient maps show subglacial features in Antarctica, long before the continent was discovered in 1820? Why have so many high ranking American and Russian officials been visiting Antarctica since 2016? Why does almost every culture have myths of a great flood and a destroyed original homeland from which their people had to relocate? Why do prophecies warn us of a future catastrophe in which the stars fall, the earth is ruined, humanity is almost completely destroyed, and a new heaven and a new earth are created? A long periodic cycle of pole shift catastrophes recurring at regular intervals explains all these mysteries and more.

Introduction

Most people do not believe that catastrophes like pole shifts happen, because there has been geological stability for thousands of years since the beginning of recorded history. Few want to consider that a great catastrophe could have erased almost all the evidence of earlier history... We also have a "normalcy bias" that makes it easy to assume that if the last few thousand years have been geologically stable then it always will be. Most people assume that human civilization has been on a track of upwards, linear progress from the Stone Age to the Space Age – and that the future will see the same stability and progress as the recent past. The same normalcy bias often destroys stock market investors, who want to believe that many years of a rising market mean it will continue to rise forever – when in fact there are known boom and bust cycles that eventually crush those who aren't watching for signs that the party is ending soon.

Geological normalcy bias was delivered to us in the 19th century with the Age of Reason. Early geologists like Georges Cuvier noticed that layers of sedimentary rock had distinct boundaries; that fossils showed certain animals were widespread and then suddenly extinct; and that lands often suddenly shifted above and below sea level, and in and out of ice age conditions. He became an early champion of catastrophism: the theory that the Earth's surface features are best explained by a series of short-lived catastrophes that cause major changes in short periods of time. Such conclusions matched stories of biblical catastrophes like Noah's Flood or the idea that the Earth was created a relatively short time ago, just before the beginning of history (which is effectively true once we realize our ancestors were not talking about the planet – which is billions of years old – and were "merely" talking about the most recent re-creation of current surface features – which are just thousands of years old.)

I have no problem acknowledging that the planet is 4.5 billion years old while simultaneously accepting that the Earth's current surface was created just thousands of years ago. This is not Orwellian "doublespeak." Ancient writers did not view our planet and "the Earth" as interchangeable synonyms. Not that long ago "terrific" used to mean "terrifying" and "gay" used to mean "happy." Such recent examples of linguistic evolution help demonstrate that words that meant one thing in a different age and language and culture occasionally suffer a translation that fails to convey the original meaning accurately.

Another example of such translation issues portrays Jesus as a carpenter. I never heard any American or English readers question that Jesus was a carpenter – but when I discussed Him with Israelis and Palestinians, they said: "Look around (I was in Nazareth doing research for an earlier book) – do you see a lot of trees? Everything here was built out of stone."

Where I live wood is plentiful. Pennsylvania means "Penn's Woods." We build wooden homes. The same goes for England, where the King James Bible was translated – the main construction material is wood. I can't blame an Englishman for translating "craftsman" as the only kind he knew – a carpenter. But wood is not the main construction material in Israel. In the original Greek text of the Bible Jesus is referred to as a "tekton" (τεκτων) which means "builder" or "craftsman" – probably a stonemason. Is the craftsman = carpenter idea another false interpretation Americans have gotten stuck with because of poor translation?

Did Jesus ever use wood or carpentry metaphors in his parables or sermons? He repeatedly uses stone metaphors – He changes Simon's name to Peter (stone is "petra" in Greek) and says to him in Matthew 16:18 "you are Peter, and upon this rock I will build My church." Jesus Himself is referred to as a stone, such as "a living stone which has been rejected," (1 Peter 2:4) or "the stone which the builders rejected, this became the chief corner stone," (Matthew 21:42) or "I am laying in Zion a stone, a tested stone, a costly cornerstone for the foundation." (Isaiah 28:16) These are references a stonemason would use. In Nazareth, Bethlehem, and Jerusalem they politely suggest that they portray Jesus as a carpenter because that is what American tourists expect – and pay for images of – but that in the Middle East He has always been viewed as a stonemason.

Likewise, when we read about "the Earth" in ancient texts, we must understand that to our distant ancestors, this phrase was not meant the way we are inclined to interpret it today. It only represented the "new" surface conditions (and the new positions of the stars in the sky) since the last pole shift – "the Earth" was a temporary configuration that meant only the current version of the earth. Consider the phrasing in 2 Peter 3:7 which contrasts "the world at that time" which was destroyed by flood, with the destruction of "the present heavens and earth." Or how about 2 Esdras 3:6, when Ezra speaks of God bringing Adam into a garden in Eden that had already existed before "the earth" was created: "And you led him into the garden that your right hand planted before the earth appeared." That garden must have been started on a previous version of the planet's surface conditions, if it had been "planted before the [new configuration of the] earth appeared." The creation of "the Earth" – as they used the term – was not billions or millions but mere thousands of years ago, just as the Bible and other myths from around the world suggest.

Unfortunately, just as ancient cultures tell us that a pole shift re-created a new version of the earth thousands of years ago – they also warn us it will happen again in the future. Chapter 24 of the Book of Isaiah is even more clear than Peter: "Behold, the Lord lays the earth waste, devastates it, distorts its surface and scatters its inhabitants... The earth will be completely laid waste... the inhabitants of the earth are burned, and few men are left... the windows above are opened, and the foundations of the earth shake. The earth is broken asunder, the earth is split through, the earth is shaken violently. The earth reels to and fro like a drunkard and it totters like a shack..." When prophecies in the Book of Revelation warn or promise that we will have "a new heaven and a new earth" I think it means there will be another pole shift. The "human measurements" and "angelic measurements" in Revelation's Chapter 21 are among the many biblical clues to the timing and extent of the coming pole shift. And I believe the Book of Genesis is describing a somewhat recent re-creation of "the earth" that occurred with the last pole shift.

Common translations say Genesis 1:1 tells us "In the beginning God created the heavens and the earth." But Isaiah 45:18 tells us "For thus says the Lord, who created the heavens (He is the God who formed the earth and made it, He established it and did not create it a waste place.)" This very clearly says that God "did not create it a waste place." Some other translations include "not empty He prepared it" and "he did not create it to be empty, but formed it to be inhabited." Yet Genesis 1:2 tells us "The earth was formless and void." If the planet was not created as an empty wasteland, how did it get that way just one sentence into the creation story in Genesis?

One top expert on the ancient meaning of Genesis is Arthur Custance. He studied the Book of Genesis in Hebrew and Aramaic before it was corrupted by numerous translations through Greek, Latin, and other languages, and reached us its final but slightly altered form in English. Custance says that in the earliest Aramaic version, Genesis 1:2 should be translated as "the earth was laid waste." He wrote: "After studying the problem for some thirty years and after reading everything I could lay my hands on pro and con and after accumulating in my own library some 300 commentaries on Genesis, the earliest being dated 1670, I am persuaded that there is, on the basis of the evidence, far more reason to translate Genesis 1:2 as 'But the earth had become a ruin and a desolation, etc.' than there is for any of the conventional translations."

God's past methods seem to involve pole shifts that cause world-destruction and renewal. With this in mind, Custance's conclusion makes perfect sense: that the correct English translation of the very first words of the Bible in Genesis 1:1-2 should be "IN A FORMER STATE GOD PERFECTED THE HEAVENS AND EARTH; BUT THE EARTH HAD BECOME A DEVASTATED RUIN."[3]

This understanding of the opening lines in Genesis is reinforced by the Edfu Texts, some of Egypt's most ancient records, which describe the first era of Egyptian history as "a period which started from what existed in the past… the ancient world, after having been constituted, was destroyed, and as a dead world it came to be the basis of a new period of creation which at first was the re-creation and resurrection of what once had existed in the past."[4] The same ideas are told in myths all over the world, as far away as the Pacific coast of North America, where tribes tell of a new earth built "on the ruins of a previous one which has been destroyed by flood or fire, usually accompanied by the collapse of the stone dome of the sky."[5]

By the 17th century, the King James Bible had corrupted our understanding of certain verses and soon almost everyone reading a Bible in English believed that the planet, and perhaps the entire universe, was only a few thousand years old. Bishop James Ussher calculated a creation date of October 23, 4004 B.C. and his date was widely accepted for centuries. Before modern science came up with data that seemed to disprove our misunderstanding of certain Bible verses, there was little reason to risk the wrath of the Inquisition or the Church and dispute unscientific ideas. But as science developed, revolutionary thinkers like Galileo eventually risked their lives to promote their observations and conclusions. By the late 1700's, Europe's most educated thinkers began to shift away from taking the Bible seriously as an accurate scientific or historical source. Stories of biblical creation of the world, Noah's Flood, and all other miraculous or divine interventions and catastrophes fell into disbelief.

In 1785 James Hutton began to popularize uniformitarianism: the theory that the Earth's surface features are best explained by extremely slow and gradual processes – the way continental drift works over hundreds of millions of years. The present – with no world-changing catastrophes – was seen as the key to the past. Hutton found plenty of evidence for ocean basins being uplifted into new land masses – he even felt this explained "the paradox of the soil." Something had to explain why soils hadn't all been washed away and depleted; rich new soil from rising sea beds provided an answer. But Hutton felt the process of uplifting, and every other geological process, was incredibly slow and gradual. Charles Lyell popularized Hutton's theory of uniformitarianism when he wrote <u>Principles of Geology</u> a few decades later in 1830. Lyell was a geologist but he was trained as a lawyer as well; he was very persuasive – and he convinced the world that Hutton was unbiased and scientific and that Hutton's theories should be viewed as scientific observation overcoming religious superstitions. Lyell was also a close friend of Charles Darwin, whose theory of evolution (natural selection) over long ages of time further discredited beliefs in the biblical creation story.

Ironically, the Catholic Church has recently embraced Darwinian evolution and the Big Bang, while denouncing Intelligent Design and Young Earth Creationism as pseudoscience. Pope Francis says that if we ignore the long-term natural processes of biological and cosmological evolution, "we run the risk of imagining God as a magician, with a magic wand able to do everything" instantly. It therefore seems like a double irony to me, that as the Church tries to catch up to modern scientific views, they are still left behind because the evolutionists have changed their views. Since 1972, when a pair of biologists and paleontologists (Harvard's Stephen Jay Gould, author of dozens of books including <u>Time's Arrow, Time's Cycle</u> – and Columbia's Niles Eldridge, also an author of many books, including <u>Reinventing Darwin</u>) wrote a landmark paper about their theory of "punctuated equilibria," most biologists agree that evolution is not gradual, but is concentrated in bursts of "quantum evolution" that preiodically interrupt long periods of stasis. The fossil record repeatedly shows long periods of calm stability with brief periods of geologically rapid events of species branching off, a theory now known as "punctuated equilibrium" and best explained by periodic pole shift catastrophes.

Sir Thomas Huxley once said: "To my mind there appears to be no sort of theoretical antagonism between Catastrophism and Uniformitarianism; on the contrary, it is very conceivable that catastrophes may be part and parcel of uniformity... Good timekeeping means uniformity of action. But the striking of a clock is essentially a catastrophe... we might have two schools of clock theorists, one studying the hammer and the other the pendulum."[6]

I suggest that we think of pole shifts from the perspective of a world that is more like a giant yo-yo than most people want to believe. If the yo-yo spins the same way for thousands of years in its descent down a very long string, inhabitants who live a small fraction of that time spend their whole lives in the reassuring stability of what seems like endless spinning in the same direction at the same speed. They would likely have little memory or record of the existence of the string, or the finger – or the series of jerky endings and sudden reversals that have occurred in the yo-yo's distant past. When a scientist on this hypothetical yo-yo world first notices that the coils of string unraveling from around the axis are not endless, and that "the end of the string" can be calculated – are they admired, ignored, or burned at the stake?

I don't assume everyone will take the pole shift theory seriously right away. When the idea that the dinosaurs were killed off by an asteroid impact was proposed in 1979 most geologists and paleontologists denounced the idea. In less than twenty years the theory became accepted as established fact. Right now, most uniformitarian geologists and historians and astrophysicists denounce the pole shift theory. But a few exceptions with scientific integrity stronger than their bias do question their old assumptions. This quote from an open-minded geologist stands out: "only time and the progress of future studies can tell whether we cling too tenaciously to the uniformitarian principle in our unwillingness to accept fully the rapid glacier fluctuations evidenced by radiocarbon dating."[7]

That geologist's quote acknowledges just one fact he couldn't ignore: carbon dating evidence of sudden changes in glaciers. There is a great deal more evidence of pole shifts in the geological record, and across many branches of science. There is even more evidence of such cataclysms in our monuments, myths, and legends. Many cultures have creation stories – and stories of destruction and re-creation of worlds. Unless the stories of global destruction are true, why would every culture look around the stable world they experienced and make up stories of repeated cataclysms? Why do we generally ignore these oral histories? For one thing, the governments and religions of the world encourage us to ignore the evidence.

Aside from normalcy bias and the desire for progress and a bright future – we must realize that governments and religions have every reason to suppress knowledge of cycles of destruction, and to influence us to believe in a stable future. Despite all the evidence that the last pole shift (geophysical as well as magnetic) was less than 13,000 years ago, many "authorities" have said that magnetic pole shifts have not occurred in the last 780,000 years, and that a crustal displacement (a rotational, geophysical pole shift) has never occurred. Yet at the same time they (<u>National Geographic</u> in this case) also recently stated that "climate change seems to be shifting Earth's geographic poles."[8] I think it is hypocritical to say geographical pole shifts have never happened, yet support their own political agenda by saying it is happening now while placing the blame on climate change!

If your government knows that a civilization-ending event is imminent, those in charge will want to prepare for it in secret, while the general population keeps on working and producing the supplies they need – not panicking, quitting their jobs, and upsetting whatever plans and facilities may exist for the survival of a select few. Since the government funds most research, most researchers feel strong pressure to have their conclusions conform. Radical and unwanted new ideas often lead to loss of funding and employment – hardly the best atmosphere for new perspectives or scientific breakthroughs.

If a major religious organization knows that such cycles are real and that a civilization-ending, end of the world as we know it event is imminent, they will want to hide evidence that discredits their version of past and future events, because they would lose power over the members of their church, and lose the donations that pay for their church. Just as the U.S. government did not amass great power by publicizing everything they do, the Roman Catholic Church did not amass great wealth and power by promoting evidence that contradicts their Holy Bible. If anything, the Vatican hides and suppresses findings that raise interesting questions. Islamic authorities have the same view of ancient artifacts and evidence in Egypt and the Middle East – they hide anything that could dispute the worldview in their Holy Qu'ran.

I found an interesting commentary on such topics at poleshiftnews.com in the article: "World Governments Don't Give Notice of Cataclysmic Events"[9]

"Will the world leaders tell you if there is a cataclysmic event about to happen? I would like to believe that they would, but the reality is most likely you will know when it is too late …during the time that it is happening. I believe the Governments would try for as long as they could to keep the system running for as long as possible... if a situation is unavoidable and likely to be a cataclysmic event of large magnitude… your Governments are not going to warn you in advance, because simply they would not want to cause panic and disruption, the other thing is how can any Government be sure of the outcome…. I remember those two Scientist's at NASA that mailed me some time ago with regard to the magnetic pole reversal currently taking place, and both virtually said the same thing, "we are aware of the situation but we are not going to make a formal announcement one way or the other as we are not sure whether it will be a cataclysmic event or a minor event."

Their uncertainty may stem from the lack of awareness of a conclusive link between a magnetic pole shift and a rotational/axial/geophysical pole shift – a crustal displacement in which the surface of the planet sees all continents and oceans change latitude and longitude simultaneously. On January 26, 2018 an article was published in Undark Magazine with the title: "The Magnetic Field is

Shifting. The Poles May Flip. This Could Get Bad." It cited data from the European Space Agency and the University of Colorado and claimed that a magnetic pole shift may be under way, and that some "parts of the planet will become uninhabitable." NASA's Bruce Jakosky, also working with the University of Colorado, had made similar warnings in 2016, as can be found in the article: "NASA Warn That Earth Is Soon Going To Experience A Full Pole Shift: Some Experts Say That Magnetic Pole Shifts In The Past Have Caused Mass Extinction Events."[10] There was also an article in Scientific American on April 8, 2016: "NASA: Earth's poles are tipping thanks to climate change." I think placing the blame on climate change is politically driven pseudoscience – but more importantly, NASA was acknowledging that a pole shift is happening. But those articles didn't generate a flood of interest like the 2018 article in Undark. That article started an internet sensation, spawning countless new videos, articles, and discussions on magnetic pole shifts and geographic shifts of the axis of rotation.

Within days, many authorities felt the need to address the sudden surge of pole shift fears, and by the end of January, Newsweek put out an article titled: "Earth's Magnetic Poles Show Signs They're About to Flip" in which they said: "the planet's magnetic field is at long last showing signs of shifting" – and they seem to attribute those words to [someone at] NASA. National Geographic downplayed what appears to be the slow start of a magnetic pole shift in an article titled: "No, We're Not All Doomed by Earth's Magnetic Field Flip."[11] They acknowledged that we have been overdue for a geomagnetic reversal and that some data suggests a magnetic pole shift is "imminent." But they make no connection to the possibility of an axial pole shift.

If I thought that the coming pole shift was "merely" going to be magnetic, I wouldn't be worried about billions of people dying in a civilization-ending catastrophe. A magnetic pole shift would still be a problem, as it would affect GPS systems and many other modern technologies – but it wouldn't end civilization or cause mass extinctions by itself. Unfortunately, much evidence suggests that, for some strange reason, mass extinctions often happen at the same time as magnetic pole shifts.

The reason seems to be that catastrophic changes in the axis of rotation seem to be linked to magnetic pole shifts. The evidence from history, geology, mythology, and many other sources couldn't be more clear. Magnetic pole shifts in the past have been found to coincide with volcanoes, earthquakes, floods, and mass extinctions. At poleshiftnews.com they wrote: "…we also end up with a rotational shift, you might ask how we at pole shift news have come to this conclusion and it's simple really, we don't take the fact that there is a rotational axis, an arctic climate and a magnetic pole all in the same regions for granted and pass them off as merely just coincidences… We believe that the magnetic poles are the main planetary mechanics, and when they change so does everything else."[12]

Although I agree with their assessment that the magnetic and geophysical poles are linked, and that if the magnetic pole experiences significant movement, we can expect a rotational pole shift to follow – the opinions expressed on a fringe or conspiracy type of web site are not always facts supported by science. In this case, however, there is plenty of scientific proof. Neil D. Opdyke wrote an article in the Journal of Geophysical Research ("Paleomagnetism, polar wandering, and the rejuvenation of crustal mobility") which lists several conclusions in his abstract worth quoting here:

"The decade from 1951 to 1961 witnessed the birth of a new geophysical subdicipline, paleomagnetism. Early studies in Europe, North America, and Australia led to the following conclusions: (1) rocks could preserve directions of magnetiziation for hundreds of millions of years… (2) …lavas had directions of magnetiziation that led to the conclusion that the mean geomagnetic field was a geocentric dipole aligned along the axis of rotation, (3) rocks… yield directions which depart widely from the present axis of rotation, (4) if these directions are used to calculate pole positions, then poles for older and older rocks fall farther and farther from the present pole of rotation, …the distribution of climatic indicators show that the pole of rotation of Earth and the paleomagnetic pole for the same periods coincide."[13] While these conclusions could also support much slower continental drift and do not prove there are rapid crustal displacements, they certainly support the idea that the Earth's crust has an axis of rotation which consistently matches the axis of the geomagnetic field. If one moves, the other follows.

As another PhD has noted on previous magnetic pole shifts: "It can hardly be maintained that enormous shifts of the position of the magnetic pole (up to 80 degrees), such as are indicated here, could have taken place while the position of the geographic pole was unchanged."[14] So as our North Magnetic Pole races away from Canada and into Russia at such an accelerating pace that the United States government stopped allowing public access to their data – perhaps we should wonder when, not if, the geographic pole shift will follow suit and catch up to the magnetic pole.

Recent evidence indicates that a magnetic pole shift may already be underway. We know that the magnetic North Pole is racing towards Siberia at a rate of at least forty miles per year, as it had accelerated to that level before the government stopped providing updates. I assume it could be migrating even faster now. Several web sites have claimed: "The magnetic north pole has moved 161 miles in 6 months" but every video and article linked as a source has been deleted. One web site I've looked at says: "A Pole Shift caused the Ice Age... government is trying to keep the earth changes at bay... the elite know about the coming changes..."[15]

I'm not assuming everything that makes me wonder is a government conspiracy to hide the truth, but I'm reminded of this quote from Fox Mulder, the main character on TV's The X-Files: "What do I do? I am a key figure in an ongoing government charade; a plot to conceal the truth ...a global conspiracy actually, with key players at the highest levels of power ...so of course no one believes me. I am ...shouting at the heavens or to anyone who will listen that 'the sky is falling, and when it hits it's gonna be the biggest shitstorm of all time!" Maybe it's not that bad – but the government did stop allowing public access to their data on the movement of the magnetic North Pole... Let's just say it might be moving substantially more than forty miles a year now.

The Earth's magnetic field overall is also getting very weak. Magnetic field strength peaked over 3,000 years ago, and the Center for Geomagnetism in Kyoto, Japan estimates that for thousands of years, (until approximately the eighteenth century) field strength only fell roughly a quarter per cent every fifty years. By the 19th century the magnetic field decline started to accelerate, and it was weakening at approximately two per cent every fifty years. By 2010, the European Space Agency said their satellites showed that magnetic field strength was falling ten times faster than previously estimated, having accelerated to about five per cent per decade. Even if the weakening magnetic field stabilized its rate of decline at that level, it wouldn't take long for catastrophic damage to our biosphere to occur – since the magnetic field protects us from cosmic radiation. Yet there was still almost no coverage of this potentially extinction level event news.

In 2014, the ESA's swarm satellite mission manager, Rune Floberghagen, emphasized the significance of the declining field strength, noted that the magnetic North Pole was moving towards Siberia, and also said one possible reason for the accelerated magnetic field decline is that: "Earth's magnetic poles are getting ready to flip." Prior to his comments, there was almost no news coverage of this, but after his comments in mid-2014 many articles suddenly recognized the importance of a magnetic field decreasing by five per cent per decade and blasted his comments all over the internet. I found one article that says: "What once took decades is now happening every year. Keep in mind, even the rate of acceleration is accelerating…" That wording sounds exactly how I would explain it! A minute later I realized – that is how I explained it. Many articles online are copying huge sections of an article I wrote years ago: "Pole Shift – Magnetic Poles First (Happening Now) Catastrophic Axis Shift Follows" without crediting my original article.

At first scientists generally said any future magnetic pole shift would take thousands of years. Then a study by Occidental said it seems like it could happen in as little as 150 years. In 2014 scientists at Berkeley said evidence seems to show it could happen in as little as 80 years. Right now I'm looking at the USGS web site (United States Geological Survey) and it says: "recent research indicates that at least one reversal could have taken place over a period of one year."[16]

And we already have the South Atlantic Anomaly – a zone of sub-crustal reversed polarity stretching from Chile to Zimbabwe. Polarity here has already reversed. As one article puts it: "If we were to use a compass deep under southern Africa, we would see that in this unusual patch south actually points north."[17] Another writer said "The Earth's magnetic field is so discombobulated over South Africa that some scientists believe we're seeing the opening strains of a planet-wide polarity change."[18] The increase in cosmic radiation over this anomaly is so severe that NASA has reported crashed laptops on space shuttle missions, and sometimes even complete satellite failure, when orbiting over this region. The SAA may or may not be a sign of a new magnetic field slowly emerging before the old one even crashes. Scientists at Rome's Istituto Nazionale de Geofisica suggest in a 2016 article titled: "The South Atlantic Anomaly: The Key for a Possible Geomagnetic Reversal" that: "there is rapid decay of the dipolar field… characteristic during the preparation of a geomagnetic transition" and that "several studies associate this anomaly as an indicator of an upcoming geomagnetic transition, such as an excursion or reversal… the rate of decay is similar to that given by previous documented geomagnetic reversals…"

Every five years, the National Oceanic and Atmospheric Administration (NOAA) updates their data on the Earth's magnetic field. They provide a forecast for the next five years so that GPS systems and other navigation and communications technology can function accurately. Wikipedia summarizes this: "The World Magnetic Model (WMM) is a large spatial-scale representation of the Earth's magnetic field… the magnetic potential of the geomagnetic main field generated in the Earth's core… predicting the temporal evolution of the field over the upcoming five-year epoch. WMM is the standard geomagnetic model of the United States Department of Defense (DoD), the Ministry of Defence (United Kingdom), the North Atlantic Treaty Organization (NATO), and the World Hydrographic Office (WHO) navigation and attitude/heading reference. It is also used widely in civilian navigation systems. For example, WMM is pre-installed in Android and iOS devices to correct for the magnetic declination. The WMM is produced by the U.S. National Geophysical Data Center (NGDC) in collaboration with the British Geological Survey (BGS). The model, associated software, and documentation are distributed by the National Geophysical Data Center (NGDC) on behalf of National Geospatial-Intelligence Agency (NGA). Updated model coefficients are released at 5-year intervals, with the current model (WMM2015) expiring on December 31, 2019."[19]

But the Earth's magnetic field has started changing so rapidly that their five year model is no longer valid. On March 21, 2018, the NOAA released a "WMM2015 Degraded Performance Notice" and a "Message to World Magnetic Model (WMM) Users" and said: "This is to inform users that the WMM Gridded Variation error has recently exceeded the performance specification in the Arctic region..." [The Magnetic North Pole is moving faster than their models predicted, to the point where the data from their broken prediction model is no longer good enough for accurate use.] "This performance degradation is caused by fast-changing core flows in the North Polar region of the Earth's outer core."[20] So the U.S. Government, which we can expect to cautiously understate the facts, is acknowledging significant acceleration of a magnetic pole shift underway. Accelerating changes in our magnetic field have literally veered off the charts of their prediction models.

But I expect far more than a mere magnetic pole shift; I expect it to be cataclysmic geophysical pole shift – a crustal displacement of the surface of the earth over the core and its axis of rotation, simply because there is so much evidence of past cataclysms. Fascinating evidence of an upcoming cataclysm that could be shared, unfortunately, is withheld from us. We, the common herd, the enemy of the state's plans to achieve certain goals using some portion of our productive efforts for as long as we can be kept working in ignorance, do not need to know. Of course not all evidence of geophysical pole shifts is denied – some is merely distorted and downplayed...

For one example, NPR did a radio show on this subject which I found in their article titled: "Scientists Solve Mystery of Earth's Shifting Poles." They began with: "Did you know that Earth's solid exterior can move around over its core, causing the planet's poles to wander back and forth? Adam Maloof, associate professor of geosciences at Princeton University, discusses the consequences of these shifts, and what may be causing them."[21] (Sounds great so far.)

"LICHTMAN: Let's talk about the terminology first. We're not talking about the magnetic poles, right?... MALOOF: We're talking about the rest of the Earth: the crust, the rest of the lithosphere and the entire mantle sliding over the outer core. So the way you imagine this is the core of the Earth, the outer part, is actually fluid iron, and it has about the viscosity of water. So we're literally sliding, you know, 2,700 kilometers of mantle over this so that, as perceived from space, what you'd see is the spin axis is staying the same, but all the continents are moving together to a new location." (Still sounds like a good acknowledgement of the facts...)

"LICHTMAN: This is SCIENCE FRIDAY, and I'm Flora Lichtman. We're talking this hour about Earth's wandering poles. Apparently, they don't stay in the same place. My guest is Adam Maloof. He's an associate professor of geosciences at Princeton University. And before the break, you were telling us that - about this huge, colossal slip and slide that happened..." MALOOF: "...we saw evidence that Earth seemed to have a shift in the poles relative to the continents on the order of 40 to 50 degrees."

These are the kind of comments I love to find! Until they say: "Some scientists think that a shift of this actually happened about 800 million years ago." So much for a realistic admission on timing. Instead he just offered a major denial of much more recent evidence of the last several pole shifts. Will he at least admit such catastrophes happen suddenly? Nope. He says: "this colossal slip and slide... [happens] at about 10 centimeters per year... maybe a little faster than your fingernail grows... we're guessing [it took] somewhere between 10 and 20 million years." Seriously?

Oh well, the article had me excited for a little while before they started to deny the evidence and play down the facts so no one gets worried. I'll choose to go with the assessment of everyone from the Egyptian priesthood to Albert Einstein that pole shifts have happened four times in the last fifty thousand years instead of that 800 million year nonsense that completely fails to explain anything from mammoths to myths to magnetism...

But despite attempts to downplay the truth, evidence continues to mount. Our magnetic field is changing so rapidly that in February 2018 the idea that a magnetic pole shift might be starting was one of the most popular online topics. In late March 2018 there are huge cracks opening along East Africa's Great Rift in Kenya. In addition to the signs that a pole shift could be developing in our very near future, other curious evidence from past pole shifts like the last one that happened almost 13,000 years ago keeps on coming - if you care to look for it.

Modern mitochondrial DNA analysis of populations around the world show a sudden unusual immigration into both Egypt and parts of North America about 12-10,000 B.C. When Egyptian mummies are analyzed, not only are their genetics interesting, but two uniquely American narcotics – nicotine and cocaine – have been found in almost every Egyptian mummy. Mummification existed on both sides of the Atlantic, as did ancient pyramids incorporating pi and other mathematical ratios into their measurements. These and other mysteries could be explained by refugees from a parent civilization like the "mythical" Atlantis.

Could the most important survivors of the last pole shift be the far-flung sailors of a destroyed civilization who were trading and colonizing around the world when destruction fell upon the earth? Better fed, better educated, and better equipped with technological innovations, such sailors would have been viewed as big, strong, healthy wise men – perhaps even as gods – by the more primitive natives of many other lands.

Imagine if a pole shift happened today, and a surviving crew from an American submarine came to shore to live with natives in the Amazon or another primitive area. Tall, strong, powerful, and knowledgeable, the crewmen would find local wives and try to teach the natives as much as possible, to rebuild from the ashes of the last civilization. The stories about them might resemble Genesis 6:4 "The Nephilim were on the earth in those days, and also afterward, when the sons of God came to the daughters of men, and they bore children to them. Those were the mighty men of old, the men of renown." Possibly, the men from Atlantis...

Our main source of information on Atlantis is Plato, who told us it was destroyed in the Atlantic Ocean over 9,000 years before his time. In the past, scholars have generally discredited his account of Atlantis as fiction because there was no evidence of a submerged continent in the Atlantic or of any civilization prior to about 3,100 B.C. One recent author surprised me by writing: "Modern historians know that no civilizations of any kind existed at that time, and it is equally clear, from the description given by Plato.... The simple and most logical solution is that the figure represents nothing but an illustration of the Egyptian's complete lack of understanding of historical timescales."[22] Is this statement merely foolishness, or deliberate deception?

To say that the Egyptians, respected throughout the ancient world for their wisdom, could have a "complete lack of understanding" of anything so basic only discredits the discreditor... Especially when coupled with the discovery of Gobekli Tepe in 1994 – a temple complex in Turkey established to be over 11,000 years old. This was well known years before the book cited above, as seen in the title of this 2008 article in Smithsonian Magazine: "Gobekli Tepe: The World's First Temple?: Predating Stonehenge by 6,000 years, Turkey's Stunning Gobekli Tepe upends the conventional view of the rise of civilization." Of course, when you are trying to discredit the truth, you might still say things like: "Modern historians know that no civilizations of any kind existed at that time." Hmmmm.

Another common misunderstanding hinders acceptance that a global catastrophe may have destroyed a very real Atlantis in what ancient Greeks called "the Atlantic." But to them the term meant the true world ocean beyond the relatively small Mediterranean Sea. Our modern concept of "the Atlantic" is a small portion of the world ocean between the Americas on the west and Europe and Africa on the east. But in ancient thinking, "the Atlantic" included all the oceans, even what we now call the Pacific Ocean and the Indian Ocean.

Plato's description of Atlantis in the middle of the Atlantic does not mean we should only look for evidence in the North Atlantic. There is a lost continent in the middle of the world ocean called Antarctica; and there is plenty of modern evidence to show that the Western portion closest to South America was warm and habitable before the last pole shift. Ancient maps show subglacial features of Antarctica, despite the fact that Antarctica was only (re)discovered in 1820. The evidence from ancient maps alone should rewrite history – every student should be taught to question orthodox history in light of such evidence   but I believe our government and religious leaders do not want us thinking about such things...

Eventually, truth has a way of coming to light no matter what. There were native cultures in the Americas with warnings of cycles of destruction which the Inquisition did not completely destroy. There are religions like Hinduism which teach us about long cycles of world ages. Around the world, "legends describe the occurrence of global catastrophes that devastated populations worldwide and precipitated a complete collapse of civilization."[23]

Many scholars analyze ancient stories and look for some kind of hidden "Bible Code" or "Mayan Code" or other encryption pattern that will reveal all the ancient messages clearly. But the truth is that ancient wise men already used anti-cryptography in their stories – incorporating the keys into the stories to make understanding them as likely as possible. These keys are largely astronomical. Some call the Bible "the greatest astrological work ever written" but in this respect it is no different from other ancient texts from around the world.

"Maya deities represent astronomical objects such as stars and planets, their activities thus describe astronomical processes."[24] Mayan astronomers based their religion, their calendar, and their monuments on astronomical alignments – alignments that occur in the early 21st century! Consider the Sphinx in Egypt; it is known for asking a riddle, and a sphinx is known to be comprised of the hind of a bull, the front of a lion, the wings of an eagle, and the head of a human. The four "animal" parts represent the four cardinal constellations of Taurus, Leo, Scorpio (often depicted not as a scorpion but as an eagle, in ancient times) and Aquarius. A major clue to the riddle (which few scholars agree on) is given by the celestial form of the riddler: look to the heavens, consider the sun's path through the zodiac.

Those who did were called priests, magi, and most importantly: astronomers. Only those who observe the heavens can understand the cosmic cycles of catastrophes, only God's giant sky-clock can measure time accurately over thousands of years, and only astronomers can read that clock to calculate when events have occurred in the past and when they might occur again in the future. In the Hindu Zatapatha-Brahmana, Manu is warned of the coming flood; he is told the year it will come and he has time to prepare a ship. In Sumeria's Epic of Gilgamesh Ea warns Utnapishtim and he builds an ark in preparation for the appointed time. In the Bible's Book of Genesis Noah is warned of a coming flood and is told to prepare his ark. In the Greek version

Prometheus warned his son Deucalion of an upcoming flood and he prepared a ship. Aside from major governments, is anyone preparing today?

If anyone were going to warn us about an approaching cosmic catastrophe today, it would be the astronomers. But unlike the ancient past, when the astronomers were the priests – now the astronomers work for the government. Most will not bite the hand that funds them. They may not be inclined to warn the common people, even when they stumble onto evidence.

I assume it was the astronomer-priests responsible for such warnings thousands of years ago. With one eye on the ancient messages left by survivors of previous cataclysms, and the other on the signs they were told to watch for in the sky, ancient astronomer-priests might successfully prepare people for a future disaster at the right time.

If we emulate them and attempt to do the same thing – look at ancient warnings and the sky – then we must consider the zodiac's symbols, and we will notice that only two signs for the constellations incorporate arrows – Sagittarius and Scorpius. The arrow of the archer and the stinger of the scorpion practically touch each other and both point towards the galactic center. What happens there that deserves our attention? Babylonian boundary stones were often carved into scorpion-tailed archer-men – a combination of features of Sagittarius and Scorpio. Did the sun between them represent a boundary between world ages?

My research has led me to conclude that pole shifts happen approximately every 12,960 years – one half precession cycle apart. I suspect that almost 26,000 years ago, one full cycle back, a pole shift occurred when the winter solstice sun aligned with the galactic center between Sagittarius and Scorpio. I suspect that the most recent pole shift occurred almost 13,000 years ago, with the summer solstice sun at the same point in the sky. (This is a northern hemisphere perspective. We should just view the conjunction of a solstice sun with the galactic center as an important astronomical clue, while realizing that it is simultaneously the winter solstice in one hemisphere and the summer solstice in the other, and would be recorded differently by different cultures in different hemispheres.) And I suspect that the next pole shift catastrophe will occur when the winter solstice sun aligns with the galactic center. It does so for a brief window of time, for a few decades every 26,000 years. We are in that crisis-window of a few decades right now.

The Mayan Long Count calendar that ended in December 2012 seems to be pointing to astronomical alignments in the time period of our early 21st century, as is their Pyramid of Kukulkan in Chichen Itza. Egypt's Great Pyramid in Giza also points to a catastrophe in the distant past and another in our very near future. Unless this was entirely symbolic, we don't have too many years left for the present world age.

Clues to consider the zodiac and certain astronomical alignments within it permeate almost every ancient story in every culture. Even the four apostles associated with Christian gospels have the same zodiac affiliations – Matthew – Ox, Mark – Lion, Luke – Eagle, John – Water Bearer. In Babylonian myth, the high god Marduk shoots an arrow through the heart of Tiamat, the cosmic Leviathan – and re-creates heaven and earth from her divided body. In the Babylonian Epic of Gilgamesh, Enkidu slays the Bull of Heaven and throws its hindquarters at the goddess Ishtar. The Hebrew judge Samson tears a lion in half with his bare hands and throws the jawbone of an ass to slay a thousand enemies. These are references to the Milky Way, Taurus, Leo, Scorpio, and Aquarius. Throughout ancient sources of hidden meaning, the anti-cryptography is really quite simple: "As above, so below." When we read about impossible heroes doing impossible things that can't really happen on Earth, we should know that the characters are really astronomical bodies, and that their actions in the stories clue us in to the timing of the events described in the sky...

More and more scientists are starting to take this seriously. Los Alamos National Laboratory's Bruce Masse, (co-author of Myth and Geology) in a 2004 article with the main title: "Myth and Catastrophic Reality" focused on South American myths – but the following quote is true of all the world's mythology: "Major natural catastrophes (e.g., 'universal' floods, fire, darkness, and sky falling down) are prominently reflected in traditional South American creation myths, cosmology, religion, and worldview. We are now beginning to recognize that cosmogonic myths represent a rich and largely untapped data set concerning the most dramatic natural events and processes experienced by each cultural group during the past several thousand years. Observational details regarding specific catastrophes are encoded in myth storylines, typically cast in terms of supernatural characters and actions." Subhash Kak said: "ancient myths encode a vast and complex body of astronomical knowledge"[25] describing ancient events in the sky while personifying the astronomical characters. I want to emphasize that cultures all around the world survived the last catastrophic pole shift and they have left us details in hundreds of similar legends, many of which we will cover later in this book.

As scientific advances are made in many fields, more and more evidence of Earth's cataclysmic past is revealed. More scientists understand, and a few are willing to commit professional career suicide and publicize their findings. A growing stream of evidence makes us reevaluate old references to earth-shaking catastrophes and "myths" of characters like Phaethon or Nibiru causing havoc in ancient times.

Nibiru may trigger the pole shift, but Nibiru is not what you have been led to believe. There is so much false information on the internet about "Nibiru" as a rogue planet that will pass through our solar system that it is difficult to look up the phrase "pole shift" without "Nibiru" dominating the search results. Right now, when I search Youtube.com for new "pole shift" videos in the last week, the first SIXTEEN results for "pole shift" all mention Nibiru. The two ideas are already so thoroughly intertwined that I must address them together.

You may be wondering: "So what? If Nibiru triggers the pole shift like you already acknowledged, why do you have a problem with Nibiru's domination of videos and articles on pole shifts?" I don't have a problem with what Nibiru really is. But the 21st century understanding of Nibiru is all wrong – pseudoscience at best, and possibly based on intentional disinformation. I suspect this is done to discredit the pole shift hypothesis, and discredit the idea that the next catastrophe could be due in the early 21st century, intentionally drawing attention away from scientifically valid pole shift evidence and replacing it with nonsense about Nibiru, instead of what the Sumerians originally meant.

The truth about Nibiru and the coming pole shift should be sensational enough. But when the lies and failed predictions about Nibiru prove to be nonsense, who will still care enough to search for the truth about the coming catastrophe? (If you are one of the exceptional few who care to understand the coming pole shift, read on...)

The meaning of the ancient word "Nibiru" has been mangled and transformed and sensationalized in my lifetime. Zecharia Sitchin made absurd claims in books like <u>The 12th Planet</u> that only he could accurately decipher ancient texts on the subject. In ancient Sumerian texts, "Nibiru" originally meant a "crossing point" like an intersection or a doorway or gateway. Root words include "eberu" (to cross) and "nibiri" (the toll one pays the ferryman for crossing the river.) Nibiru was associated with the galactic center at the intersection of two great lines in the sky - the spine of the Milky Way, and the plane of the ecliptic – and the high price we pay when the solstice sun crosses it.

The Maya viewed the dark cleft at the center of the Milky Way as a gateway and symbolized it with a door. It could be "the open door which no one can shut" in Revelation 3:8 and may also be the location in the Song of Solomon 2:14 "in the clefts of the rock, in the secret place of the steep pathway, let me see your form." Those verses may not clearly indicate the celestial throne is in the center of the Milky Way, but Revelation 4:6-7 says: "Before the throne there was something like a sea of glass, like crystal; and in the center and around the throne, four living creatures full of eyes in front and behind. The first creature was like a lion, and the second creature like a calf, and the third creature had a face like that of a man, and the fourth creature was like a flying eagle." The sea of glass could be the dense field of stars near the central bulge of the Milky Way, and the four creatures are definitely the four cardinal zodiac constellations of Leo, Taurus, Aquarius, and Scorpio (often represented as an eagle or a scorpion.) The throne of God is described as being in the middle of the zodiac in the sea of crystal...

I believe many ancient cultures viewed the galactic center this way, as the suddenly illuminated throne/seat of God – especially during its active phases every 12,960 years or so when the galactic center becomes brightly visible. Such a divine timetable could explain why "ancient theology was a science based on number" in so many cultures.[26] Research from astrophysicists, from Carl Seyfert to Paul LaViolette, suggests there are "tremendously energetic explosions that periodically take place at the centers of galaxies."[27] All spiral galaxies, including our own Milky Way galaxy, have a galactic center that periodically spews out matter and energy. For tens of thousands of years after the initial outburst at our galactic center, a wave front with an electromagnetic pulse followed by a gravity wave and different wavelengths of light and other radiation spreads across the galaxy. "The front of the wave would have a very steep gravitational gradient that would be capable of exerting a very strong inward pull on any planet or star it passed."[28] This galactic superwave eventually reaches our solar system, becoming visible – even in daytime – as a bright blue light. It becomes visible almost at the exact moment it triggers a pole shift.

Before this event horizon reaches us and makes the galactic center visible, the location of this throne of God is only known from ancient history passing on the secret. The Sumerian creator god Enki was often hidden and was believed to hibernate for long periods of time. The supreme high god of the Maya, the hidden god Hunab Ku, was responsible for all movement and measure in the universe but remained invisible. The Inca worshipped the pre-Inca supreme god Pacha Camac, who was invisible despite being the one almighty god, maker of heaven and earth. India has the hidden god of fire, Agni, whose divine spark gives the fire of life and whose divine will powers all that happens in the universe. Egypt's high god Amun (or Amon or Amen), who was so prominent that many of us still invoke his name today at the end of our prayers (Amen) had a name meaning "the hidden one" or "invisible."

But eventually the cycle of precession brings the solstice sun (Ra, or Re) into conjunction with the galactic center and we have the composite god Amun-Ra. Then when the galactic center becomes visible as a blue light it turns into the Eye of Horus. Our ancient ancestors in other nations also saw this; it is India's Navel of Vishnu, the Hopi's Blue Star Kachina, The Mayan Heart of Heaven, the Arab's Tariq Star, the Vikings Odin's Eye (at the Well of Mimir), the Aztec's Place of Light Where He-Who-Gives-Light Hides Himself... and the Sumerian's NIBIRU. When it appears, we have a problem.

Quadrillions of tons of ice accumulate near the North and South Poles, but not precisely centered at the poles. The Antarctic ice sheet's center of mass is over 300 miles off center today. Additional mass imbalances exist underneath the surface. This puts significant and ever increasing torque on the crust of the earth, nudging the off-center ice mass towards the equator. But the magnetic field of the earth generates high viscosity (resistance to flow) in otherwise liquid molten iron and lava, and the friction between the layers of the earth keeps everything in place. Friction prevents the crust from gradually adjusting to the torque of these tangential centripetal forces and stops the crust from rotating away to correct the imbalance. The imperfect shape of the Earth, with an equatorial bulge, also stabilizes the surface. But if the earth's magnetic field weakens enough, it would be like turning off an electromagnetic lock – the friction between molten layers under the crust would disappear. We would suddenly experience a pole shift and the surface of the planet would quickly and catastrophically change.

The last time this happened was probably just over 12,900 years ago... At that time Nibiru - the crossing point of the ecliptic and the Milky Way - was clearly illuminated by a bright new blue star - the suddenly visible, periodically active galactic center spewing out all sorts of dust and light and radiation including high levels of gamma radiation and more importantly – a gravity wave.

A noteworthy astronomical alignment at this point in the sky existed when the summer solstice sun and Jupiter and the active galactic center came together, and their visible alignment near each other may have appeared to cause the pole shift – although "modern findings indicate that these catastrophes were triggered by intense volleys of cosmic ray particles [and incoming dust and a gravity wave] originating from the core of our galaxy."[29] Many myths describe a gathering of the gods to decide mankind's fate – probably an alignment with many additional planets visible within a few degrees of the galactic center, the solstice sun, and Jupiter...

Zeus has been identified with both Jupiter and the galactic center, and in Plato's description of the destruction of Atlantis, in the very last sentence of Critias, "Zeus called together all the other gods to judge the fate of Atlantis, 'and when he assembled them, he spake thus:....'"[30] but then Plato says nothing else after this apparent astronomical alignment at the crucial time of civilization-ending judgment. The Egyptians described their sun-god Ra, riding in the Boat of Millions of Years, when he reached the Great Palace, described like a city at the center of the universe, where he "entered at the head of his holy mariners and established himself on the throne of the two horizons." Another legend says that Ra gathered the gods in secret to judge mankind. "Let them all come hither in secret so that men may not behold them, and fearing, take sudden flight. Let all the gods assemble in my great temple." These holy mariners or other gods are presumably the visible planets, seemingly assembled for judgment of mankind at the great temple, the throne in heaven. This crossing point of the ecliptic and Milky Way is the crossing point at the galactic center where a very bright blue light occasionally appears at the time of a great catastrophe. Sumerians called it "Nibiru."

Over time, the blue "star" of Nibiru fades away... and the memory of what happened, and the astronomical bodies that aligned thousands of years ago are no longer remembered or understood. By Babylonian times, they still thought of Nibiru as the highest god's star, at a crossing point, splitting the sky in half and ruling over the rotation of the heavens, setting the proper motion of the stars... They also associated it with Jupiter, but only sometimes...

Despite what you may read on the internet, or in sensational books like Zecharia Sitchin's The 12th Planet – Nibiru is not "Planet X" and there is no rogue planet that almost crashes into Earth every 3,600 years. Nibiru was the appearance of the suddenly active galactic center as a new blue "star" at the crossing point of the ecliptic and the Milky Way. It is mentioned often because it appeared to be the avatar of the Wrath of God at the time of a cataclysmic pole shift that devastated the surface of the planet. If you want to understand what Nibiru originally meant, read a translation of the ancient Babylonian/Sumerian creation story: The Enuma Elish. Read about pole shifts, the galactic center, Carl Seyfert, Paul LaViolette, the Book of Revelation, and ancient mythology... but not Zecharia Sitchin's The 12th Planet.

If there were a 3,600 year cycle of destruction – then there would have to have been great destruction no more than 3,600 years ago. But well-established Egyptian history goes back over 5,100 years, to the unification of

Upper and Lower Egypt into one kingdom. Even the biblical chronology for the Great Flood (which I suspect leaves many centuries unaccounted for and yields too recent a date) suggests that Noah used his ark at least 4,300-4,400 years ago.

If there is no 3,600 year cycle – and no approaching Planet X to warn us of imminent danger as it comes closer – then it may be much more difficult to calculate when the next pole shift will occur. We may have to use clues found across many disciplines – geology, astronomy, history, mythology, and prophecy may all need to be studied to contribute to our understanding. But the evidence does exist, however scattered the pieces may be. If we are up to the challenge of reassembling the pieces of the puzzle, we may see the final picture.

Evidence of Previous Pole Shifts

For pole shifts to occur, we must assume that the outer layers of the Earth's crust are not always firmly attached to the layers beneath them. The surface of the Earth must occasionally detach itself from what lies below and move as one solid piece over the core, like the chocolate around a cherry. While the central majority of the planet keeps rotating normally deep underground, the surface we live on must suddenly move in a new direction. If this happens slowly and gradually over geological ages we would barely notice the slow drift. But if it happens suddenly, then tsunamis and earthquakes far larger than any in modern history would reinvent entirely new surface conditions.

Polar regions would move towards the equator, and formerly warm areas would move towards the poles. The ice sheets of Greenland and Antarctica would melt and raise sea levels for years to come while other areas, now at the poles, would enter a new "ice age." The oceans and the atmosphere, currently rotating West to East with the rest of the planet, would initially keep their momentum, and crash into coastlines with mile-high tsunamis and hurricane style winds. The Earth's crust would buckle like an accordion and rise to new heights where land previously spread out over the equatorial bulge is forced to fit into a slightly smaller area away from the bulge. Other lands approaching the equatorial bulge would crack and spread apart over their new larger surface area, causing features like the Great Rift that runs a sixth of the way across the world, from the Dead Sea in Israel south to the great African lakes. Such stretching and buckling and cracking forms mountains, causes earthquakes, and unleashes a hellfire of volcanic lava and ash. Few people would survive the end of our current world to tell myths about how there had been a civilization capable of great things...

I once started a fictional story with: "I loved my grandfather's stories about America. It was like he had been there himself, living in one of its impossibly huge cities with wealth and inventions beyond imagination in the days when they made dreams happen. His arms waved with emotion describing various heroes and accomplishments and how in the end it was all destroyed in a single day. Of course none of us believed that Strong Arm had ever thrown men to the moon or that One Cup had discovered how to make suns on the ground or that America had ever really existed... Or maybe we did want to believe; maybe that is why I became fascinated with it in the first place. Maybe that is why I was fired from the Hall of Records and I'm on trial for heresy today."

Is such a story really that different than the legends left over from the Maya or the Aztecs or other Native Americans about their ancestors and original homelands? Is it that different than the epic tales from ancient India, Egypt, or Greece about the distant past? Or the stories of the creation – and re-creation – of the world over many ages? Is it so different from biblical stories about being kicked out of Eden? The main difference as I see it is that my story about America as an unbelievable myth takes place in a fictional future. For the myths of many cultures, similar stories are simply what was able to be passed down as history...

Of course, the idea that catastrophic pole shifts repeatedly destroy civilization and alter the entire surface arrangement of the Earth – this is just a theory. It is not a universally accepted fact. And for theories to gain credibility, supporters must provide a great deal of evidence.

Some of the best evidence for pole shifts includes:

Bands of coral

Wherever there are shallow oceans near the Earth's equator, there are tropical coral reefs. The water must be shallow – no more than about 200 feet deep – or the algae Zooxanthellae with which the corals live in a symbiotic relationship don't get enough light for photosynthesis. Corals can live anywhere within about 30 degrees north or south of the equator where the water stays above 16 degrees Centigrade, (61 F) but various species are very sensitive to heat and can only survive in a narrow temperature range of a few degrees. Some thrive in the hottest tropical waters, others in cooler seas with seasonal variations a bit farther from the equator. Think of it as distinct stripes of coral species running parallel with the equator. If the surface of the Earth stayed fixed in its current orientation for millions of years, there would only be one circle of coral reefs around our planet at our current equator.

But instead there are many bands of ancient coral remains marking many previous equators, crisscrossing the Earth every which way. Ancient coral reefs buried in sedimentary rocks are a common source of oil in the Middle East and other regions. One ancient coral equator even bisects the Arctic Ocean. As Charles Hapgood tells us there is even "evidence of warm coralline seas stretching right across Antarctica."[31] He cited coral expert Professor Ting Ying H. Ma of the University of Fukien, China – formerly the founder of the Geology and Oceanography departments at the University of Taipei and the Chief of the Oceanographic Section of the Chinese Geographical Institute. After studying corals in Alaska, Spitzbergen, and Antarctica, Dr. Ma concluded that many total displacements of the entire outer crust of the earth must have taken place.

In Ma's own words, the evidence he found in ancient coral reefs could only be explained by "the sudden sliding of the solid earth shell over the liquid core" leading to "a sudden change in latitudinal positions."[32] The movements must have been fast, because the edges of the coral bands are not blurred and spread out as they would be if a gradual pole shift had occurred over centuries or millennia and allowed the coral to gradually grow into adjoining areas.

In his examinations of ancient coral reefs, it was clear that the width of coralline seas remained steady, that global temperatures remained steady (there were no global ice ages cooling down the entire planet and killing off sensitive coral species) – the evidence suggested that the continents repeatedly moved together as a single shell and repositioned new lands and seas within the tropical band capable of supporting coral near the equator. Such cataclysms also explained how ancient coral reefs are found at much greater depths where coral reefs never form. (There are cold-water corals at higher latitudes and at lower depths, such as off the coast of Scotland, but they form patches, not reefs like in the tropics.)

Ma's analysis of coral showed that the waters around Japan and China had a much warmer climate in the Pleistocene Age than it does today (much like Siberia was a lot warmer.) One of his much older ancient equators passed through the Ural Mountains in Russia, and through Sichuan and Yunnan provinces in China. Many times corals showed our cardinal directions were different, for example in one period waters around England and France were much warmer than Germany, indicating England and France were both closer to the equator, and that Germany, presumably, was north of both of them. Coral equators passing through what is now the Arctic Ocean, and through Antarctica, were already noted above.

Bands of Soil

The same conclusions can be drawn from bands of soil. Due to catalyzing effect of sunlight on various chemical reactions in soil and rocks, differing amounts of sunlight create different ratios of various mineral compounds at different latitudes. Chemically similar soils are formed in bands of tropical and temperate zones around the earth based on latitude – with circles of arctic soils formed at the poles. Dr. George W. Bain noted equatorial-formed soils in the arctic and circles of polar soils in equatorial regions and concluded that old climatic zones have no relation to today's equator. He found one former equator ran right through the New Siberian Islands in the Arctic Ocean. In Bain's own words the "fixity of the axis of the earth relative to the outer elastic shell just is not valid" and "the recurrent change in position of these rings through geologic time can be accounted for now only on the basis of change in the position of the elastic shell of the earth relatively to its axis of rotation."[33] Even the U.S. Geological survey notes: "rock strata containing alternating horizons of marine and non-marine fossils"[34] occur everywhere they look. Pole shifts explain these frequent and regular changes in altitude and latitude.

Changing magnetic fields in lava rock

When orange-hot liquid lava flows from a volcano, it soon cools down and solidifies. In liquid form, ions of iron and other magnetic molecules align themselves in the magnetic field of the Earth. We can analyze solid lava rock and determine which direction the ions point towards, and it will indicate both the direction of the closest magnetic pole, and the distance to it – because the declination of the ions is almost vertical near a pole and close to horizontal if the lava cooled near the equator. We can also tell that one layer of lava rock may have come long before or after another if vast changes in the direction of the embedded magnetic ions exist between layers. The earliest rock magnetization analysis I have seen was published in 1955 by John W. Graham, and after studying ancient rocks his conclusion was "the rocks were magnetized by a geomagnetic field about like the one today, the essential difference being that this field was in a significantly different orientation. This result is discussed in terms of slip of an outer shell of the earth relative to the axis of revolution."[35]

Japanese and Soviet studies of such lava rock ionization are quite extensive and show dozens of magnetic North Pole locations (not necessary matching the rotational North Poles precisely) throughout the Pleistocene Age.[36] Solidified minerals and volcanic rock on the seafloor near the expanding Mid-Atlantic Ridge show alternating stripes of magnetic polarities in different directions. But great variation in the magnetic field within the same sample is most interesting, for that lava solidified during a pole shift.

Lava fields at Steen's Mountain in Oregon show north itself was moving up to six degrees per day while the lava cooled. One citation says: "eight degrees per day"[37] but this could be an error. The original papers submitted by Robert Coe and Michel Prevot in 1989 under the title "Evidence Suggesting Extremely Rapid Field Variation During a Geomagnetic Reversal" claimed at least three degrees of change per day, and that the magnetic field changes during solidification "implies astonishingly high rates of change of the geomagnetic field." (In later years, with additional samples analyzed, they were less cautious and admitted movement of six degrees per day.)

Their most likely explanation was "rapid variation of the geomagnetic field direction during cooling." Making very conservative estimates and assuming the slowest possible cooling, they conclude that the entire event must have happened "within 15 days. This period is undoubtedly an overestimate... Nonetheless, even this conservative figure of 15 days corresponds to an astonishingly rapid rate of variation of the geomagnetic field direction of at least 3 degrees per day." (At six degrees per day, this only takes a week.)

Some would argue that this evidence merely implies a magnetic pole shift, and not necessarily a change in the rotation of the planet's crust over the core. Some argue there is no such thing as a "merely magnetic pole shift" because a magnetic shift is a precursor to a crustal displacement over the core. Others point out there is no way to tell the difference – an apparent movement of the magnetic pole is either caused by the magnetic pole moving below the crust, or by the crust moving over the magnetic core. Peter Warlow suggested the latter argument in his article about pole-flipping in Britain's <u>New Scientist Magazine</u> in November 1978.

Warlow noted that most scientists accept that magnetic reversals and magnetic pole shifts have happened countless times, but they doubt that any shift of geography or any apparent change in the axis of rotation follows suit. But Warlow says of his theory: "Ironically, perhaps, and seemingly contradictory, one of its essential features is not to have a magnetic reversal at all. Instead I propose a geographic reversal" changing the alignment of the Earth's magnetic field "not by a direct action on the field itself, but indirectly through a geographic reversal." If the crust of the planet shifts over the core, the core's magnetic field may stay the same, but how it appears to us on the surface, where new lands move over the magnetic poles, would seem as if the magnetic poles had moved. As for the duration of a pole shift in Warlow's opinion: "I suggest that the actual reversal takes place in a matter of days… [Expect] massive tidal waves, carrying vast quantities of debris and sediment from the ocean floors and depositing them over the land… augmented by the equally inevitable volcanic activity… Some fauna, large or small, could be rendered extinct…"

As for the views of Coe and his colleagues analyzing the lava at Steen's Mountain, they also seem to suggest the pole shift recorded in the lava was not merely a magnetic pole shift, but that there was significant movement between layers near the surface of the Earth. Several pages into the paper the authors note that "fluid velocities near the core-mantle boundary… indicated by our data could imply much higher velocities on the order of 1 km/hr."[38]

If the crust in this location moved even one kilometer per hour, (which they again considered a conservative estimate) that still means that this location (far from the pole which would be experiencing even greater movement) over 15 days could move 360 kilometers over the core. There is not enough information to know how significant the change in location was for the site in what is now Oregon, but we can safely assume that it moved at least a few hundred miles in relation to the core of the Earth, that it saw a change of at least thirty degrees orientation in the magnetic field, and that the movement of the poles, which change MORE than any other parts of the crust, experienced even greater change.

Coe and his colleagues, even though they had minimized their findings with very conservative estimates, were still soon put under pressure from their peers in mainstream academia to renounce their findings. There was no known terrestrial mechanism capable of initiating such a pole shift, so many scientists assumed the data must be wrong. The idea that the magnetic field and or the crust of the earth may have moved more than three degrees per day seemed to support biblical creationists, not establishment scientists, and that kind of controversial dispute with mainstream ideas puts careers in jeopardy.

Steen's Mountain was revisited, and the lava reanalyzed. Over 75 samples were taken from over 50 different basalt lava flows. The conclusions, published in "Nature" in 1995 (Volume 374) reported less conservative, even more amazing results: "the rate at which the orientation of the ancient geomagnetic field had rotated could have reached an astonishing six degrees per day over an eight day period."

Let that sink in. That's up to 48 degrees of southward movement of the North Pole in eight days. Imagine Paris or Seattle moving south to the new equator in about a week. (And possibly submerging deep below sea level, given the miles of seawater comprising much of the equatorial bulge...) Or Rio de Janeiro, now within the tropics, moving south beyond the new Antarctic Circle and rising several miles in elevation as it leaves the equatorial bulge, freezing over and getting buried under snow and ice at its new latitude and elevation.

Coe's findings at Steen's Mountain are not an isolated aberration. Geologist Scott Bogue found lava flows at Battle Mountain, Nevada that show a change of fifty-three degrees between two flows that formed one year apart. The two flows could have been a day apart, a week apart, a month apart, or up to about one year apart. Bogue said: "We're trying to make the case that [the new work] is another record of a superfast magnetic change."[39] The evidence here does not prove a pole shift can occur within just a week or two, as in the previous example at Steen's Mountain, but it still helps prove that massive geomagnetic changes can occur in very short amounts of time.

Flash-Frozen Mammoths

The most famous mammoth carcass was found by Russian hunters near the Berezovka River in the year 1900. "The Berezovka Mammoth" had at least eleven species of unchewed grasses, mosses, beans and buttercups in its mouth. Many of these plant species do not live in Siberia today under current climate conditions. There were also undigested sedges and grasses and buttercups in its stomach. Given the known life cycle of these flowering plants, it was concluded that the mammoth died in "late July or the beginning of August" and was frozen almost instantly. That is an oddly warm time for a sudden freeze.

The ground below the mammoth did not offer any of the vegetation in the mouth or belly; the last meal was apparently interrupted by a sudden problem that made him flee from wherever he had been foraging. The ground underneath him was also soaked with his blood, so he did not drown in a river or flood. His erect penis at the time of death indicates a state of terror and freezing. Scientists struggled to find an explanation, like a fall into an icy crevasse in the permafrost during an earthquake, to explain the temperate climate foods in its mouth and stomach, along with the well-preserved carcass obviously frozen shortly after death and remaining frozen ever since. I suggest he may have been trampled and wounded as a herd of mammoths fled a suddenly flooding lowland in a stampede for higher ground.

He was wounded badly enough to stop running and die, and for the ground beneath to be soaked in his blood, but the erection shows that freezing temperatures killed him before too much blood was lost. Summer temperatures must have fallen very fast. Imagine how much body heat a mammoth must have while alive, and how quickly chewed vegetation in stomach acid would deteriorate if not quickly frozen. What kind of temperature drop must have occurred to freeze a mammoth solid so quickly? Would a sixty degree drop in two hours be enough? I suspect it would have had to be even more extreme.

Maybe one mammoth could fall into a fissure in permafrost opened during an earthquake, and be frozen in a warm climate... But there have been many dozens of well-preserved mammoths found in the last century in Siberia – and they would have rotted away within weeks if the temperatures remained much above freezing after they died. A sudden pole shift provides a better explanation. The Berezovka mammoth carcass was found near 67 degrees north – at least, in our present surface conditions. Many suspect that Arctic Siberia must not have been in the arctic when lush grasslands full of summer flowers provided food for herds of such large herbivores that may have eaten over 500 pounds each every day... yet the mammoth was frozen before decomposition set in, under compressed layers of snow...

In 1977, Siberian gold miners led Soviet scientists to the remains of a young mammoth carcass (soon named "Dima") near the headwaters of the Kolyma River, approximately 63 degrees North Latitude. It was recovered from a clear lens of pond ice, but the silty water in its lungs show it probably drowned while struggling to get out of the water. A.V.Lozhkin wrote about this mammoth in 2016, in Volume 145 of Quaternary Science Reviews... after noting that forests and other warm weather plant life both required for mammoths to survive (with adults eating hundreds of pounds of vegetation per day) and confirmed by actual plant remains found nearby, Lozhkin concludes: "the concept of mammoth-steppe needs rethinking in light of new paleobotanical data."

"Dima" was too young to eat; the digestive tract showed nothing but mother's milk... but other mammoth carcasses found nearby show a diet of sedges, willow leaves, grasses, and bog mosses – all of which were found directly under Dima's body. Yes, the idea that mammoths lived in an icy wasteland and never ate anything seems to be a huge misunderstanding. Dr. William Farrand wrote in an article titled "Frozen Mammoths and Modern Geology"[40] that "an apparent paradox remains – that the climate in northern Siberia was warmer than at present at some period in late glacial time when climates elsewhere on the earth were cooler than at present."

Only a pole shift – a displacement of the entire crust of the earth – explains how one region gets much warmer and another gets much colder at the same time. This is what Ivan Sanderson concluded in "Riddle of the Frozen Giants," in the Saturday Evening Post in which he suggested that "the part of the earth where their corpses are found today was somewhere else in warmer latitudes at the time of their death."[41] Yet they were quickly frozen, with eyeballs and skin and stomach contents frozen intact and remained frozen ever since. Hmmmm.

Sanderson wrote on similar topics many times, including an article in Pursuit in 1969 in which he notes that: "In the New Siberian Islands... whole trees have turned up; and trees of the family that includes the plum; and with their leaves and [ripe] fruits. No such hardwood trees grow today anywhere within two thousand miles of those islands. Therefore the climate must have been very much different when they got buried... nor could they have retained their foliage if they were washed far north by currents from warmer climes... either what is now the Arctic was at the time as warm as Oregon, or the land that now lies therein was at that time elsewhere. Geophysicists don't go for an overall warming of the planet to allow such growth at 72 degrees north; otherwise everything in the tropics would have boiled! ...the whole earth's crust shifted."

In 1886 the Arctic explorer Baron Eduard von Toll found a frozen fruit tree ninety feet tall with green leaves and ripe fruit still on the branches – surely the source of Sanderson's quote above – and surely only possible if the climate changed drastically within a single day, remaining frozen ever since. Otherwise, such a tall tree would not have remained fresh before being entombed in snow and ice and remaining so for thousands of years.

One Soviet researcher, I. P. Tolmachev, noted that mammoth remains are usually found near the highest elevations of the tundra. He also noted that many are found in frozen mud – mud which must have been unfrozen and warm when the animals stepped into it. Since such animals are not stupid, we may speculate that there was a major distraction forcing their rapid movement into unfamiliar but warm high terrain. Since muddy silt does not accumulate on high ground, we might also assume that a flood brought the muddy silt just before the region froze.[42] The dense concentration of hundreds of thousands of mammoth remains on the New Siberian Islands may be explained by the animals' forced retreat to the highest peaks as the poles shifted and the seas rose. What would you do if you noticed floodwaters rising, possibly at a peak rate of hundreds of feet per hour? You would get to high ground, and hope you were high enough to survive.

In the November 1968 issue of Sputnik, the Soviet publication "reported the discovery of evidence of human occupation of the New Siberian Islands, as well as Spitzbergen, during the ice age." The New Siberian Islands are currently only 10-12 degrees south of the North Pole. Snow covers the coasts from mid-September to early June, and the average temperature in January is -29 degrees Centigrade, or -21 degrees Fahrenheit. Wikipedia says that "during cold years, islands may remain ice-blocked through the summer." Yet the Soviet newspaper Kommunist Tajikistana gave additional details of human habitation there, describing "a Stone Age settlement on the Novosibirsk Islands" [New Siberian Islands] with "bone implements and arrowheads, as well as needles and axes" and "prehistoric cliff drawings... [with] well-preserved incised outlines of whales and deer" on Spitzbergen.[43] There is no way human settlements could have existed there with the latitude and climate the islands have today.

In 2011 British scientists and BBC cameramen reached an even better preserved mammoth carcass named "Yuka" found in the northern coastal area of Russia's Yakutsk region, near Ust Kuyga, approximately 70 degrees North Latitude. Temperatures were close to negative 60 degrees Centigrade (-76 F) when they reached the almost complete carcass of a juvenile mammoth covered in strawberry blonde fur that showed signs of both being attacked by lions, and being stolen away from lions by humans, immediately after being killed. While many mammoth remains are dated back tens of thousands of years and many mammoths like "Dima" didn't necessarily die during the last pole shift, "Yuka" is believed to have died approximately 10,000 B.C.

Neither mammoths nor lions nor humans were likely to have thrived in this part of Siberia with its present temperatures. And one additional piece of evidence suggests that mammoths were not well-suited to cold climates: the lack of sebaceous glands in the skin. Most animals living in cold climates have sebaceous glands to secrete oil and lubricate the skin to prevent dryness and cracking in the cold, dry air. Microscopic studies of mammoth skin have shown that mammoth cells and skin layers are absolutely identical to the skin of Indian elephants today. "The lack of oil glands in the skin of both animals... showed a negative adaptation to cold."[44] Polar bears have sebaceous glands in their skin. Humans have them too, concentrated in the face. But mammoths and elephants have never had them.

Kyle Bennett wrote an article (based on Chapter 5 of a book he at least planned to write in 2011, though I don't see it for sale under this title: <u>Polar Wandering and the Cycle of Ages</u>) called "Mammoths of the Last Polar Age,"[45] in which he said:

"Russian scientists undertook an investigation in 2001, in collaboration with NASA... They investigated the climate of Faddeyevskiy Island, within an archipelago called the New Siberian Islands. This island is located at 72° North... with a mean annual temperature of −15°C. ...The thickness of the permafrost in some parts of the island is 400 to 500 metres.

...These Russian scientists analyzed ancient pollen and plant macro-fossils... Their results showed that Faddeyevskiy Island was not covered by ice... [and] nearby Kotel'nyy Island was free of ice between the LGM [Last Glacial Maximum] and the beginning of the Holocene [our current Age.] The flora of Faddeyevskiy Island during the last Ice Age included grasses and sedges, and even roses, buttercups and daisies. These conditions made the island a suitable home for many creatures, with enough vegetation to feed populations of grazing mammals...

So how could horses and mammoths live on the New Siberian Islands during the very height of the North American Ice Age? This question is simply avoided by the geologists who have studied these islands. It doesn't fit in with the dogmas of mainstream science, so it is simply ignored. The main reason it is ignored is because this evidence utterly destroys the idea of a global ice age, involving a lowering of global temperatures..." You can't say the world was cold enough for an ice age to cover North America with an ice cap down below forty degrees latitude, while parts of Siberia above seventy degrees latitude had a warm climate.

Bennett continues: "...The only plausible explanation is that the rise and fall of Ice Ages is caused by successive pole shifts, and that the North Pole was in Canada when mammoths were living on what are now island archipelagos above the Arctic Circle. But geologists prefer to live in a virtual reality, where they ignore the facts and mislead the public about the true history of Ice Ages. They are supported by the mainstream media, which perpetuates the myth of global Ice Ages and publishes propaganda designed to marginalize and discredit the ideas of Earth Crust Displacement...

To paraphrase and adapt the words of Anthony Browne (from his book The Retreat of Reason), there is a conspiracy not so much of silence but of denial that stretches across the media and mainstream Science. There is endemic dishonesty towards the public, but because everyone is in denial... few realize it.... This received wisdom is in fact easy to disprove – it just requires them to look at the facts – but... This collective denial so envelopes the media-scientific elite that they have little idea how detached their world-view is from reality.

The presence of a warmer Ice Age climate in the Arctic than today is irrefutable, but when it comes to discussing the cause of Ice Ages, this fact ceases to exist. Mainstream geologists then develop their theories based on the assumption of a global Ice Age, wherein the Arctic was colder in the Ice Age. So they are practitioners of doublethink – when it suits them, they can believe that black is really white."[46]

The number of mammoth remains found in Siberia actually increases the farther north you go. The New Siberian Islands, in the Arctic Ocean, are often referred to as an elephant or mammoth graveyard. Entire islands seem to be made of little else other than a frozen muck of twisted mammoth bones and tusks, shattered tree trucks, and sand and silt. Thousands of tusks, and tens of thousands of mammoth bones simply fall into the Arctic Ocean every year. It begs the questions: how did so many mammoths ever live at these latitudes, and what killed them off?

Hugh Brown suggests: "The mammoths' graveyards can be considered as additional evidence of the recurrent cataclysms of the earth. Their shallow burials make it appear probable that they lived... when the Hudson Bay Basin was at the North Pole... An exhibit at the American Museum of Natural History showing a similar group of skeletons of prehistoric animals, all piled together like offal at a slaughterhouse, can be explained most rationally by the deluge caused by a careen of the globe. Those animals evidently came to their death by cataclysmic mass drownings."[47]

Professor Frank Hibben made very relevant remarks commenting on similar mass graves of North American animals found in Alaska and other areas. He wrote in The Lost Americans: "In many places the Alaskan muck is packed with animal bones and debris in trainload lots. Bones of mammoth, mastodon, several kinds of bison, horses, wolves, bears, and lions tell a story… frozen solid, lie the twisted parts of animals and trees… It looks as though in the midst of some cataclysmic catastrophe of ten thousand years ago the whole Alaskan world of living animals was suddenly frozen in mid-motion…"[48]

Many chapters later Hibben wrote: "The Pleistocene period ended in death. This was no ordinary extinction of a vague geological period which fizzled to an uncertain end. This death was catastrophic and all-inclusive… What caused the extinction of forty million animals? …Whole herds of animals were apparently killed together, overcome by some common power… We have gained from the muck pits of the Yukon Valley a picture of quick extinction… Neither the Pleistocene animals nor their untimely end are phenomena peculiar to the American continents. Asia was deeply involved."[49]

He continues later: "Whole herds of animals were apparently killed together, overcome by some common power… Mingled in these frozen masses are the remains of many thousands of animals killed in their prime… The animals were simply torn apart and scattered over the landscape like things of straw and string, even though some of them weighed several tons. Mixed with the piles of bones are trees, also twisted and torn and piled in tangled groups; and the whole is covered with fine sifting muck, then frozen solid."[50] The obvious conclusions are that mass animal graveyards are caused by a sudden tsunami flood which quickly freezes, and that ice ages are not caused by global cooling but by a region's sudden movement into a polar latitude.

"Ice Ages" are definitely evidence of pole shifts

Aside from the pole shift theory we are discussing, in which the crust of the planet experiences a cataclysmic reorientation over the core, with all continents and oceans changing their latitude and longitude simultaneously – "no one has ever explained the cause of any Ice age or the cause of the end of this ice age. It certainly did not just happen by itself."[51] Robert Felix, author of the iceagenow web site and several books on ice ages and magnetic reversals, has realized part of the connection... He suggests that the cause of ice ages "has to do with magnetic reversals... ice ages have been returning in a fairly predictable cycle tied to the precession of the equinoxes... Magnetic reversals... trigger ice ages and that they return in a cycle, a regular pattern that we should actually be able to predict." Felix also noted that "these things happen tremendously rapidly... Ice Ages correlate with magnetic reversals."[52]

I am glad that he makes these connections, but it is only partial progress. To me this seems like looking at a crime scene and blaming a weapon without thinking one step further to find the criminal that is the ultimate cause. Many researchers notice that magnetic pole shifts are linked to ice ages but fail to look a step further and realize the link to crustal displacements.

Pole shifts involving a change in the position of the crust of the planet relative to the core of the earth and the rotational axis are by far the best explanation for ice ages at various locations throughout history; an area experiences an ice age when it is at one of the poles. Ignoring the pole shift explanation, "scores of methods of accounting for ice ages have been proposed, and probably no other geological problem has been so seriously discussed, not only by glaciologists, but by meteorologists and biologists; yet no theory is generally accepted."[53]

"Ice Ages," as described by mainstream science, are periods of time during which the Earth experiences unusually cold temperatures and continent-sized glacial ice sheets expand south farther from the poles than usual. The problem with this explanation is that they have no idea why the ice sheets expand southwards at one position, but have no ice sheet at other locations at the same latitude. For example, in the last Ice Age, the Pleistocene, which ended (for no conventional known reason) approximately 12,900-11,700 years ago saw the Laurentide Ice Sheet cover all of northeastern North America, dipping as far south as Missouri, Tennessee, and Kentucky – as low as approximately 36 degrees latitude. What is now New York City was under ice approximately a mile deep.

Yet in Siberia, there was no ice sheet. Not at 36 degrees north, not at even at 40 or 50 or 60 or 70 degrees North. Large animals like mammoths, rhinoceroses, horses, camels, and bison (many of which need to eat hundreds of pounds of vegetation each day) roamed Siberia in the tens of millions in a temperate climate in lands now at latitudes up to 72 degrees north. If the Earth had entered a climactic phase of unusually cold temperatures, why was it only unusually cold in what is now Canada and the Eastern United States, and warm in what is now Russian Siberia? "The former grassy flora of late Pleistocene Siberia has been compared with today's African savannah"[54] and it supported an almost identical population of large herbivores and carnivores. The mainstream explanations of ice ages admit a lack of understanding of how this could happen.

The pole shift theory explains it easily: the North Pole was in Hudson Bay at the time. Eastern Canada was covered in an ice sheet because it was at the North Pole. It had an "Ice Age" in the same way Greenland and Antarctica have them now – they are located at the poles now. When the Laurentide Ice Sheet formed, evidence shows it did not spread south from our current North Pole – the ice spread out in all directions from Hudson Bay. (The Bay itself is there because the weight of the ice sheet depressed the crust of the Earth and formed a basin underneath it.)

The Aztec capital city of Teotihuacan, "the place of the gods," is aligned 15.26 degrees east of north, along an axis through the Avenue of the Dead which, if extended far enough, crosses Hudson Bay. Was the site established before the last pole shift? Anthony Aveni, author of dozens of books including Skywatchers of Ancient Mexico, wrote that of 56 Meso-American sites he studied, 50 of them are curiously oriented east of north. In the Old World, many sites like the Temple Mount in Jerusalem and the Ziggurat of Ur are oriented west of north. Are they originally aligned with a previous North Pole because the sites were established long before the current orientation of the globe? Thomas Mills wrote a book called Stonehenge: If This Were East in which he analyzes the positions of the stones and concludes that the arrangement only makes sense if it were aligned to the sky when the North Pole was in Hudson Bay. On a related note, Professor David Bowen applied Chlorine-36 rock-exposure dating to the stones of Stonehenge to determine when they were first exposed to the air – and determined they might be up to 14,000 years old.[55]

Ancient monuments also clue us in to an even more ancient North Pole location. Baalbek, a site in Lebanon with the largest stone blocks in the world, some weighing over 400 tons – seems to be aligned properly if the North Pole were in the Norwegian Sea east of Greenland – just where it probably was two pole shifts back, at least 25,900 years ago. Look up "ancient equator" on Youtube and many videos will show that many of the most impressive ancient monuments and temple sites can be found on a single line that circles the earth, a potential former equator that crosses the Egyptian Pyramids at Giza, Petra, Persepolis, Mohenjo-Daro, Angkor Wat, Easter Island, the Nazca lines and Macchu Picchu in Peru, and many other sacred temple sites. We should also note that Macchu Picchu and Angkor Wat are exactly opposite each other, as are Mohenjo Daro and Easter Island, with interesting ratios and distances between several of the other sites on the line as well. Modern architecture surely has been built over these very ancient sacred sites, for if this line is an old equator, then the North Pole was near the border of southeast Alaska and the Yukon at the time – and researchers like Charles Hapgood conclude that was THREE pole shifts back, at least 39,000 years ago.

There is evidence of "ice ages" in India where the ice sheet existed all the way to our present equator. Huge glacial debris beds in Northern India show that an ice sheet started in the south, closer to our current equator, and spread northwards. But the ice sheet was local, just like the one centered on Hudson Bay; the ice did not cover the entire planet down to the equator (and we would have little surface life left on Earth if it had – let alone tropical species which obviously survived "India's Ice Age" somewhere warmer. In South Africa there is evidence of glaciers but they spread north to south – starting closer to our present equator and spreading away to our south. Central Africa also shows evidence of an ancient ice sheet centered on the Sudan Basin. Yet tropical life survived... Either ice sheets have started near the equator and moved north and south from it – or the lands under the ancient ice sheets were not equatorial, but polar, when the ice sheets formed. I think pole shifts are a much better explanation.

A.P. Coleman wrote a book in 1929 titled: <u>Ice Ages Recent and Ancient</u>, in which he said: "In early times it was assumed that during the glacial period a vast ice cap radiated from the North Pole, extending varying distances southward over seas and continents. It was presently found, however, that some northern countries were never covered by ice and that in reality there were several more or less distinct ice sheets starting from local centers, and expanding in all directions, north as well as east and west and south. It was found, too, that these ice sheets were distributed in what seemed a capricious manner. Siberia, now including some of the coldest parts of the world, was not covered, and the same was true of most of Alaska and the Yukon Territory in Canada; while northern Europe, with its relatively mild climate, was buried under ice as far south as London and Berlin; and most of Canada and the United States were covered, the ice reaching as far south as Cincinnati in the Mississippi Valley."[56]

Coleman focused on the United States and Europe, but also noted that one ice cap in India centered on a spot at what is presently only 17 degrees north latitude, and would have covered a huge [now] tropical area including part of the equator. He even noted the evidence of a South African writer on the same subject, who focused on the southern hemisphere, and found that aside from the present time and the ice cap in Antarctica today, "the ice caps of all geological periods in the Southern Hemisphere were eccentric as regards the South Pole, just as the Pleistocene ice caps were eccentric with regard to the North Pole."[57] This was not even explicable due to high altitudes; Coleman's evidence found ice caps had existed at sea level in what are now tropical areas in Asia, Africa, and Australia. The pole shift theory explains the locations of these ice caps, and why they were "eccentric to the poles." They weren't centered on our current poles, but on former ones. "Either we accept that the Antarctic ice cap is the *first* continent sized ice sheet *ever* to have been situated at a pole… or we are obliged to suppose that earth-crust displacement" has repeatedly changed the positions of former ice sheets.[58]

We have plenty of evidence that the North American ice cap centered on Hudson Bay is not the only ice cap with an inexplicable (without the pole shift explanation, anyway) location. But because this last one was so recent, with evidence so close to American universities who study such things, we know the most about this ice cap. Dr. Fred Earll, a geology professor at Montana College of Mineral Science and Technology, even noticed an issue with the way this ice cap melted: "the ice sheet seems to have melted from west to east, and from north to south, as well as from south to north… how can this be reconciled with our assumption…" that the climate was warm in the south but not the north?

There are two assumptions, really, neither of which explain the melting patterns as well as a very fast pole shift bringing Hudson Bay and its ice cap to their current location. If the "assumption" were the orthodox/standard one, the idea that there hasn't been any pole shift at all, then not only should the Laurentide Ice Cap have been centered at the present North Pole, rather than on Hudson Bay – but it should have formed first and been thickest in the north, and it should have been thinnest and melted first in the south.

In correspondence with Professor Hapgood, however, Earll asked "how can this be reconciled with our assumption that the pole was moving [very slowly over centuries or millennia] at this time from Hudson Bay to its present site in the Arctic Ocean[?]" The ice should not have melted from north to south under the conditions of slow polar shift either. The ice in the north of the ice cap should have melted last. If the North Pole slowly progressed northward underneath it before leaving the northern edge of the ice cap after many centuries, the ice should have melted last at the northern edge of the former ice cap. But it didn't happen that way. The ice cap melted simultaneously starting at all its edges and the middle disappeared last. Hapgood wrote: "Considerable thought on this matter had led me to the conclusion that the evidence points to a very rapid transit of the pole from its old to its new home… If the transit of the pole was very sudden, then the pattern of melting indicated by the evidence becomes understandable."[59]

Continent-sized ice sheets have suddenly appeared and disappeared. They have done so in various areas without affecting other regions at the same modern latitude, which would seem to rule out periods of global cold spells as an explanation for ice ages. Unless and until a better explanation comes along, I must go with the simplest and most obvious one: "ice ages" are regional events that occur when a region happens to find itself near the North or South Pole after a pole shift. Ice sheets melt in a region when the next pole shift carries those lands to warmer latitudes and starts new ice sheets and ice ages somewhere else.

Mountain Building

The pole shift theory also adequately explains the formation of mountains for the first time. The equatorial bulge of the earth means there is a larger planetary radius and surface area along the equator than at the poles. When a pole shift occurs, land moving to the equator must crack open and expand to fill the greater area, perhaps about three feet per mile assuming a pole shift of thirty degrees latitude.[60] Over hundreds of miles along the meridian of movement, this could yield a total of a thousand feet of new rifts, with hundreds of small ones, or just a few new canyons opening up. Is the Mid-Atlantic rift evidence of one such prehistoric event? The larger the crack in the crust, the more likely that magma expands to fill the void from underneath, with lava pouring onto the surface as a volcano. Albert Einstein's correspondence with Charles Hapgood notes that the pole shift theory "was the only theory he had ever seen that could explain the volcanic zones."[61] Land moving away from the equatorial bulge must buckle like an accordion into a slightly smaller area. This explains why mountain ranges form in long lines. Vening Meinesz noted that "In large parts of the earth's surface... tension seems to exist in the crust [in one region] at the same time that folding takes place elsewhere, and this fact is difficult to reconcile"[62] with traditional attempts to explain mountain formation. Buckling like an accordion into mountains in one region, while simultaneously tearing open rifts in another, is exactly what the pole shift theory predicts.

One geology textbook concluded, without the benefit of the pole shift explanation, "It must be admitted, therefore, that the cause of compressive deformation in the earth's crust is one of the great mysteries of science."[63] As another confused geologist lacking a good theory of mountain building put it:

"great mountain chains suggest by their structure... episodal disturbances of such indescribable and overwhelming violence, that the imagination refuses to follow the understanding."[64]

Yet another geologist notes that "long narrow belts of parallel folds" in mountain chains like the Appalachians "call for the action of some great horizontal force thrusting in one direction... It is an explanation that explains nothing that we want to explain."[65] Columbia professor Dr. Walter Bucher acknowledged that mountain range "belts are the result of world-wide stresses that have acted on the crust as a whole. Certainly the pattern of these belts is not what one would expect from wholly independent, purely local changes in the crust."[66]

Another geologist named Umbgrove noted that "stratigraphic studies make it increasingly evident that the terrestrial crust was subjected to a periodically alternating increase and decrease of compression... I feel there is overwhelming evidence that the movements are the expression of a common, world-wide, active, and deep-seated cause."[67] Umbgrove also noted "a sort of periodicity to its operation." He said "The geologist comes across periodicity in many of the pages which he is arduously deciphering – in the sequence of the strata [marine, terrestrial, marine, terrestrial] ... and their contents of former organisms... subsidence in one area and then in another... the intrusion of liquid melts or "magma" rising... the rhythmical invasion of the continents by epicontinental seas and the subsequent retreat..."[68]

All this and more is simply explained by periodic pole shifts roughly every 12,900 years. If one is due soon, we might experience a heavy concentration of earthquakes near the equator, on opposite sides of the planet along the meridian of movement of any pole shift about to occur. I am reminded of Bible verses like Mark 13:8 suggesting that "earthquakes in various places" "are merely the beginning of birth pangs."

Yet another category of evidence in support of pole shifts:

Human Records

Almost every culture has creation stories involving the destruction of a previous world and the creation of a new world. Founders or ancestors or demigods or giants who were greater than we are today had to flee ancient homelands like Eden, or Atzlan, or Dilmun. There are monuments that align not with current cardinal directions like North or East, but with ancient locations where the North Pole was long ago, including the Sphinx Temple, the Ziggurat of Ur, the Wailing Wall in Jerusalem, the entire city of the ancient Mexican capital - Teotihuacan, Copan, Xochicalco, and many other ancient sites. I am not suggesting that the Jewish Temple or any other construction was built at least 12,900 years ago – but they seem to have been built on top of foundations from something older that were probably aligned to true north before the last pole shift.

And some structures are clearly pre-historic. The Pyramid of Cuicuilco outside Mexico City has a stone plaza around it, covered by 18 feet of sediment and a lava flow dated by multiple methods to have flowed over three sides of the pyramid approximately 6,500-7,000 years ago. Gobekli Tepe in Turkey is conservatively dated to be at least 11,000 years old. Tiahuanaco in Bolivia has been dated anywhere from 11,000 to 15,500 years old.

Perhaps most impressive of all is the "old equator" of ancient sites including Easter Island, the Nazca Lines, Machu Picchu, the Giza Pyramids, Petra, Ur, Persepolis, Mohenjo Daro, Angkor Wat, and several other sites on a circle around the Earth. Mohenjo Daro and Easter Island are exactly opposite each other, as are Nazca and Angkor Wat. This seems to indicate the North Pole was near the border of southeast Alaska and the Yukon at the time the plan for the sites of these monuments was originally made – though the monuments we see on these sites today may be newer monuments built on top of ancient sites…

In Hermetic tradition, the Egyptian wisdom god Thoth "succeed[ed] in understanding the mysteries of the heavens [and to have] revealed them by inscribing them in sacred books which he then hid here on earth, intending that they should be searched for by future generations but found only by the fully worthy."[69] Perhaps these are "books of stone" like the pyramids and other monuments that, when the measurements are analyzed, reveal an incredible amount of scientific data...

Myths and legends about ancient floods and ancestors with amazing abilities and technologies are easy to appreciate. If they are based on real events at all, they obviously and clearly describe the global destruction of at least one advanced civilization prior to recorded history as we know it. But there are other forms of evidence that require painstaking analysis to appreciate. One such category of evidence involves:

Ancient maps.

Professor Charles Hapgood, author of Earth's Shifting Crust, wrote another book called Maps of the Ancient Sea Kings: Evidence of Advanced Civilization in the Ice Age in 1966. In that book he details years of analysis studying numerous early maps, and his conclusions must alter our view of history if we pay attention to how he reached them. I think his logic is inescapable and his conclusions seem to be correct. Anyone who gets Albert Einstein to write the foreword to his book must be doing something right...

After studying dozens of mankind's oldest maps, Hapgood concluded that the map of the Turkish Admiral known as Piri Re'is, made in 1513, was made just as the admiral told us: from several ancient source maps going back to at least the time of Alexander the Great. The map not only shows Antarctica, a continent not officially (re)discovered until 1820, but "the map shows non-glacial conditions extending for a considerable distance inland."[70]

Captain Arlington Mallery had already written a book about "Lost America" in 1951 in which he noted that ancient maps of Greenland showed accurate subglacial details – mountains and rivers under the ice sheet. Mallery "discovered" the Piri Re'is map (he made its existence known to the public after recognizing what it probably shows) after a navy buddy brought it to his attention in 1953. In 1956, Georgetown University Radio interviewed him about "New and Old Discoveries in Antarctica."

At one point in their conversation, the moderator said: "Mr. Mallery, this must then lead to the conclusion that there were competent explorers and map makers along the coasts of the Atlantic long before Columbus." Mallery responded: "Several thousand years before. Not only explorers, but they must have had a very competent and far-flung hydrographic organization, because you cannot map as large a continent as Antarctica, as they apparently have – half of it – or as extensive as Greenland, or half the continent of North America, as we know they did – probably [at least] 5,000 years ago. It can't be done by any single individual or small group of explorers. It means an aggregation of skilled scientists…"

Years before Hapgood, Mallery and M.I. Walters of the U.S. Navy Hydrographic Office had already reached such conclusions about the Piri Re'is map. Hapgood contacted Mallery, and also decided to contact the cartographic staff of the U.S.A.F. Strategic Air Command, and see if they would come to similar conclusions. U.S. Air Force Lt. Col. Harold Ohlmeyer wrote back to Professor Hapgood on July 6, 1960, after analyzing the map:

"Dear Professor Hapgood,

Your request of evaluation of certain unusual features of the Piri Reis map of 1513 by this organization has been reviewed.

The claim that the lower part of the map portrays the Princess Martha Coast of Queen Maud Land, Antarctica, and the Palmer Peninsular, is reasonable. We find that this is the most logical and in all probability the correct interpretation of the map.

The geographical detail shown in the lower part of the map agrees very remarkably with the results of the seismic profile made across the top of the ice-cap by the Swedish-British Antarctic Expedition of 1949.

This indicates the coastline had been mapped before it was covered by the ice-cap.

The ice-cap in this region is now about a mile thick.

We have no idea how the data on this map can be reconciled with the supposed state of geographical knowledge in 1513.

Harold Z. Ohlmeyer Lt. Colonel, USAF Commander"

In Hapgood's <u>Maps of the Ancient Sea Kings</u> he also notes that details on the Oronteus Finaeus Map of 1531 suggest "that Antarctica was visited and perhaps settled by men when it was largely, if not entirely, non-glacial."[71] Many "experts" would say that Antarctica has been entirely covered with ice for millions of years, but we know that uniformitarian nonsense is wrong, and no one is suggesting that people mapped Antarctica millions of years ago. Since maps seem to show ancient but non-glacial features, and some evidence like the Ross Sea sediments suggest glacial conditions only completely spread over the Antarctic coastlines as recently as 4,000 B.C. – let's just say that the civilization that mapped the interior of Antarctica did so many thousands of years before we are told there were sailors who could get there.

But the Oronteus Finaeus Map of 1531 has one major flaw – the 16th century copyist apparently misunderstood (could not read) the very ancient maps he copied, because he used their circle for 80 degrees latitude as his Antarctic Circle, drawing the entire continent far too large, almost touching South America, Africa, and Australia. Adjusting the scale down, the map from 1531 is extremely accurate. Hapgood brought this map to the attention of the Air Force as well, and on August 14, 1961 the chief of the Cartographic Section of the Strategic Air Command at Westover Air Force Base, Captain Lorenzo Burroughs wrote back:

"Dear Professor Hapgood:

It is not very often that we have an opportunity to evaluate maps of ancient origin. The Piri Reis (1513) and Oronteus Finaeus (1531) maps sent to us by you, presented a delightful challenge, for it was not readily conceivable that they could be so accurate without being forged... [but] we have concluded that both of these maps were compiled from accurate original source maps, irrespective of dates...

...The agreement of the Piri Re'is Map with the seismic profile of this area made by the Norwegian-British-Swedish Expedition of 1949, supported by your solution of the grid, places beyond a reasonable doubt the conclusion that the original source maps must have been made before the present Antarctic ice cap covered the Queen Maud Land coasts.

"It is our opinion that the accuracy of the cartographic features shown in the Oronteus Finaeus Map (1531) suggests, beyond a doubt, that it was also compiled from accurate source maps of Antarctica, but in this case of the entire continent. Close examination has proved the original source maps must have been compiled at a time when the land mass and inland waterways of the continent were relatively free of ice... The Cordiform Projection used by Oronteus Finaeus suggests the use of advanced mathematics... on a stereographic or gnomonic type of projection involving the use of spherical trigonometry."[72]

The civilization that mapped Antarctica also mapped the entire world, from the Americas to China, and from Antarctica to Greenland. They knew the correct size of the Earth – far better than Eratosthenes, who had calculated its size with an error of almost five per cent. Much as our modern civilization uses different map projections suited to different portions of the Earth, the ancient mapmakers had many different ways of representing small, large, tropical, and polar areas on different maps, and many of their projections used spherical trigonometry to draw the maps – a branch of mathematics we thought had been invented recently.

The Gothenburg Magnetic Excursion and Flip

First discovered in Gothenburg, Sweden – then at other sites all over Sweden, and eventually the world – a global magnetic flip was discovered in core samples. Around 12,500 years ago (keep in mind a degree of inaccuracy – this date does not mean it happened centuries after the pole shift) the Earth's magnetic poles flipped after drifting strangely for a few decades. This does not *necessarily* mean that the rotational poles did anything – but the magnetic poles suddenly moved until magnetic North was in the central Pacific near the equator, then even more suddenly, the magnetic poles shifted to the exact opposite of what they had been before the movement started – magnetic north and south switched places. Because this magnetic pole shift coincides with other evidence, we can assume it was directly related to the catastrophes and extinctions. Many studies show a close correlation between magnetic reversals – which in and of themselves pose no threat – and mass extinctions. Some of the earliest articles on this correlation include "Evolutionary pulsations and geomagnetic polarity" written in 1965, and "Faunal extinctions and reversals of the Earth's magnetic field" from 1971. A reasonable conclusion would be that a crustal displacement occurs at the same time as the magnetic pole shifts and mass extinctions.

South American Evidence

We could describe the mass extinctions of animals in South America as well, including additional species like toxodons – which looked like a hippopotamus and a rhinoceros had interbred – which died out approximately 13,000 years ago. But 46 toxodons are carved into the stone artwork on "The Gate of the Sun" of the Bolivian city of Tiahuanaco. The stonework also has carvings of Macrauchenia, a now-extinct three-toed horse – and Cuvieronius: a recently extinct elephant-like animal with tusks and a trunk.[73] This city has been dated by some researchers like Arthur Posnansky to be approximately 15,000 years old. Age aside, it certainly used to be a massive port city with great stone docks. But it is now at an altitude of over 12,500 feet above sea level, and thirteen miles from the shore of Lake Titicaca. Some of the stonework in Tiahuanaco looks like it was suddenly abandoned before the work was finished. What happened?

Lake Titicaca, apparently, used to be part of the ocean. It still has salt water mollusks, sea horses, and much other evidence that its life and its waters were thrusted up from sea level.

The ground near Tiahuanaco drops off sharply in elevation, falling about 35 feet in altitude some distance north of the city. Above that level, the ground is rich with ancient artifacts. Below that level, only the stone rings used on fishing nets are found. The lake itself shows a clear strand line in the mountains, about 295 feet above the water line at the north end of the lake. But the strand line falls to 274 feet below the current water line of Lake Titicaca if you go far enough south. It then progressively dips even farther at a steeper incline. It would seem that the entire region, if not the whole continent, was unevenly uplifted from the ocean in one or more pole shifts.

Some ruins, structures, and terraces near Tiahuanaco can be found on Mt. Illimi as high as 18,400 feet above sea level – well above the current line of eternal snow where no one should have built or farmed anything – under current altitude and climate conditions.

The French paleontologist and disciple of Cuvier, Alcide d'Orbigny, wrote around 1832 "It would seem that one cause destroyed the terrestrial animals of South America, and that this cause is to be found in great dislocations of the ground caused by the upheaval of the Cordilleras... I argue that this destruction was caused by an invasion of the continent by water... My final conclusion from the geological facts I have observed in America is that there was a perfect coincidence between the upheaval of the Cordilleras, the destruction of the great race of animals, and the deposit of Pampas mud."[74] Charles Darwin also commented a few years later: "To destroy animals, both large and small, in Southern Patagonia, in Brazil, on the Cordillera of Peru, in North America up to Behring's Straits, we must shake the entire framework of the globe."[75]

The Usselo Horizon

First discovered in 1940 in the town of Usselo in the Netherlands, this ten-centimeter thick layer of black ash, soot, charcoal, and sand has been dated to approximately 12,700 years ago. The layer has been found not just in the Netherlands, but also in Great Britain, France, Germany, Poland, Belarus, Egypt, South Africa, India, Australia, the United States, Vietnam, and several other nations. In Europe, the Usselo layer covers artifacts from Magdelanian cultures and marks their sudden demise. In the United States the layer is usually referred to as the "black mat" overlaying Clovis culture artifacts and the remains of millions of large mammal species that died out in the Pleistocene extinction event. 95% of all mammals over 100 pounds went extinct in North America and Siberia approximately 12,900-12,700 years ago, even though frozen remains show full bellies, healthy bodies and numerous young. Millions of deaths seemed to suddenly interrupt a warm climate with well-fed, healthy populations of animals.

The evidence in Siberia is well known because finding frozen mammoth carcasses gets more attention than those of other animals. But comparable to the "mammoth horizon" in Siberia are the frozen muck deposits in Alaska that yield about 8,000 carcasses a year in the Fairbanks area alone. Almost every day, gold-digging operations find the prehistoric remains of lions, saber-toothed tigers, mammoths, rhinoceroses, giant sloths, mastodons, horses, bison, camels and dozens of other species. Dr. R. Dale Guthrie of the Institute of Arctic Biology, commenting in 1982 on the sudden demise of these animals "and other Pleistocene species, one cannot help wondering about the world in which they lived. This great diversity of species, so different from that encountered today, raises the most obvious question: is it not likely that the rest of the environment was also different?"[76] Despite the bodies being torn to pieces by surging waters, skin, hair, and all soft tissues are almost always found well-preserved, as the animals' drowned and dismembered carcasses were frozen suddenly, then buried beneath a protective layer of soot. I want to emphasize that the vegetation that supported these vast herds of animals was growing in the surrounding soil – the ground was not frozen, and the climate was not cold – before the Pleistocene mass extinction and sudden freeze.

Global weather seems to have changed instantaneously. Frank Hibben estimates that "40,000,000 animals died in North America alone."[77] Three species of lions, and three species of elephants – lived in what is now North Carolina – prior to the last pole shift. What is now the United States had animal diversity like central Africa has today. W.B. Scott wrote that "the extraordinary and inexplicable climatic revolution had a profound effect upon animal life, and occasioned or at least accompanied, the great extinctions, which, at the end of the Pleistocene, decimated the mammals over three-fifths of the earth's land surface."[78] Yet despite the quick freeze that preserves the remains of so many animals in the Arctic – most bodies are found torn apart, not intact – along with the shredded remains of trees – a testament to the extremely violent end that came to these locations along with the sudden drop in temperature.

The Pleistocene Extinction, tied to a thick layer of ash, soot, and charcoal around the world, reminds some geologists of the iridium-rich layer marking the comet impact that killed off the dinosaurs 65 million years ago, as that layer was also rich in charcoal and soot. Subsequent tests of soil from the

Usselo/black mat layer show high levels of iridium. Scientists argue over the possibility of a comet impact as the cause for both the Usselo Horizon and the end of the ice age, vs. incoming radiation and cosmic dust from a galactic core explosion – but the extraterrestrial nature of the cause of the global fires almost 13,000 years ago is rarely questioned. As Randall Carlson once said, there is an "energy paradox" in regard to what could have triggered a pole shift, and what could have quickly melted the North American ice cap, and what could have started fires around the world: "The melting was accompanied by an injection of energy into the terrestrial system that had no known source – there was no source of terrestrial energy capable of achieving that kind of effect" on the ice or on the surface of the earth.

Many researchers, including Graham Hancock – whose work I have immense respect for – are convinced there was a comet impact over North America at this time. I can admit that some evidence supports this comet impact theory. But other evidence discounts it, as summarized in the abstract for the article: "Nanodiamonds and wildfire evidence in the Usselo Horizon postdate the Allrod-Younger Dryas boundary." It states: "The controversial Younger Dryas impact hypothesis suggests that at the onset of the Younger Dryas an extraterrestrial impact over North America caused a global catastrophe. The main evidence for this impact – after the other markers proved to be neither reproducible or nor consistent with an impact – is the alleged occurrence of several nanodiamond polymorphs, including the proposed presence of lonsdaleite, a shock polymorph of diamond. We examined the Usselo soil horizon... No lonsdaleite was found... Our analysis thus provides no support for the Younger Dryas impact hypothesis." (I am left leaning towards the galactic superwave hypothesis instead, in which cosmic dust enters the solar system, causing darkness and cold on earth followed by intense solar flaring and heat.)

Another interesting layer, possibly related to the Usselo Horizon, is the last Heinrich layer in ocean sediments. Hartmut Heinrich discovered many layers of marine sediment containing little but rock grains of continental origin. The most accepted theory is that glacial ice occasionally has catastrophic sudden meltdowns in which an estimated "2.3 million cubic kilometers"[79] of glacial ice quickly release all the rocks, gravel, pebbles, and sand embedded in them as they previously grinded over the surface of the land. The last Heinrich event was dated to approximately 12,000 years ago (but may have lasted from about 12,900-11,700 years ago. This all coincides with the "sedimentary boundary (YDB) separating the Younger Dryas from the preceding Bolling-Allerod at a depth corresponding to 12,900 years before present."[80]

Waterfalls

Waterfalls may be majestic and beautiful but they are not geologically stable. There is heavy erosion where rock crumbles away at the top ledges and also at the bottom of the falls where the falling water and sediment grinds the underlying rock away. At Niagara Falls, the U.S. Geological Survey concludes that the American side erodes upstream at an average of 2.5 feet per year, and the Canadian side of the falls retreats about 4.5 feet per year. Older estimates for the age of the falls conclude they have retreated about seven miles and have only existed for approximately 8,000 years; modern conclusions suggest closer to 12,000 years – depending on the source. "The birth of the Niagara River and the record of the short span of its life history during our present epoch are proof of a recent careen of the globe and a world cataclysm."[81] Analysis of the much smaller Falls of St. Anthony on the Mississippi River near Minneapolis also yield an age of at least 8,000 years, with the modern consensus at around 11,700 years. Most waterfalls in the United States are believed to have formed from glacial melt at the end of the last "ice age" when the North American ice sheet (no longer at the pole) rapidly melted and formed new rivers.

Are Penguins and Seals Evidence of Pole Shifts?

Penguins: these black and white Antarctic birds look so adorably cute that we may forget to ask the obvious question: why would any animal (though Emperor penguins are the best example) lay its eggs 60-100 miles deep into Antarctica, during the winter months when there is no sunlight, no food, and temperatures hover around forty below zero? Could this behavior possibly have started under such harsh conditions? Or are penguin breeding habits practically proof that Antarctic weather was much warmer when penguins first formed an attachment to the land there?

I am forced to assume that the penguins formed a territorial attachment to Greater (Eastern) Antarctica at some point in the very distant past, several pole shifts back, when the latitude was farther from the pole and the climate was mild. They hardly bother with Lesser (Western) Antarctica, which was warmer and more habitable during the previous orientation of the Earth's surface. At first this seems strange, that they would have avoided the portion of Antarctica that had been warmer... where evidence in sediment off the coast shows there were at least some ice-free rivers flowing as recently as 4,000 B.C. But this makes sense if Western Antarctica was habitable to humans, and our ancestors hunted them.

Both Dr. Urry, and researchers from the University of Illinois independently found sediments off the Ross Ice Shelf and the Pacific coast of Antarctica with layers of red clay interrupted with layers of calcerous ooze and organic remains; also substantial changes in the size of pebbles and gravel, indicating a shift from flowing rivers to glaciers scraping bottom – indicating ice-free rivers flowing in Western Antarctica as recently as 6,000 years ago. If man can live in Iceland, Norway, Russia, Alaska, and Greenland today – they could have lived on ice free coasts of Western Antarctica, and eaten penguins.

If there is any doubt regarding how much the climate could have changed there... a quick comparison to the climate changes in England at the end of the last ice age should do. Wikipedia tells us: the "Bolling-Allerod interstadial was an abrupt warm..." period ending 12,700 years ago. "At the height of the Bolling-Allerod, winter temperatures in the British Isles had increased by 25 degrees centigrade... to levels typically found in that region today."[82] If winter temperatures in England rose 45 degrees Fahrenheit, that means today's average January temperature of 41 degrees had been -4 degrees prior to the end of the last ice age/pole shift. Coincidentally the warmest part of Antarctica – the Antarctica Peninsula that juts up towards South America – has an average winter temperature of approximately 0 degrees Fahrenheit – though it varies anywhere from 23 to -13 it is usually between 5 to -4 degrees Fahrenheit in winter.

Assuming that Atlantis may have been on the Antarctic Peninsula before the last pole shift, and assuming temperature changes similar but opposite to those England experienced – the climate there may have been 45 degrees warmer in winter – warmer than English winters today. The peninsula stretches from approximately 63-75 degrees south latitude today – but it may have been thirty degrees farther north if the North Pole was in Hudson Bay. 33 to 45 degrees south latitude means the climate in pre-pole-shift times may have been like the East of the United States today. The weather in cities of Antarctic Atlantis may have been like today's weather in Buenos Aires, Cape Town, Sydney, or at least as nice as in San Francisco, New York, or Tokyo.

I can understand birds laying eggs on lands at those latitudes. I believe the climate of Antarctica must have been warmer when emperor penguins first chose their breeding grounds.

Seals are another example of animal life in polar conditions that should make us question global conditions of the past. The same seals that thrive today on the coast of Alaska have stranded populations in Lake Baikal in Siberia and in the Caspian Sea that borders southern Russia and northern Iran. Despite the thousands of miles of land between these groups and the huge differences in climate, latitude, and longitude between these populations in our current surface conditions, it seems likely that in a previous position of the earth's crust prior to one of the last pole shifts all three locations with seal populations were connected and had similar climates.

Tree Rings

The word dendrochronology is derived from the Greek words for tree (dendron) time (chronos) and science (logos). Trees grow in rings, one ring in the wood trunk of the tree for each year. Some years have a good climate and are especially thick; other years with a bad climate can lead to thin rings. Using samples from the remains of many thousands of different trees, scientists have been able to establish a continuous tree ring record dating back approximately 12,500-12,600 years. If my theory that the last pole shift was approximately 12,900 years ago is correct, it would make sense that most trees were destroyed not long before that time, and that there would not be much evidence from which to obtain samples for dendrochronology until forests reestablished themselves at least a century or two later - so I would expect the farthest tree ring records would probably go back would be to about 12,600-12,800 B.C. It is also interesting that carbon dating is believed to yield accurate results back about 50,000 years, but carbon dating results cannot be corroborated that far back using wood as biological evidence. Dendrochronologically-dated tree rings are used for the period up to about 12,500 years ago. Beyond that, corroboration is derived from independently dated marine samples such as foraminifera and corals as far back as 26,000 years ago. It would seem that another wall is reached, with another shortage of coral evidence beyond 26,000 years ago - two pole shifts back – when, not coincidentally, there was also a pole shift.[83]

Pole Shifts on Other Worlds

Earth is not the only place where pole shifts happen. The largest mountain in the solar system is Olympus Mons, a volcano on the Martian equator three times the height of Mount Everest. But it wasn't always on the equator. The mass imbalance of this giant protrusion above the Martian surface is believed to have brought it there in a "geologically rapid pole shift."[84] Dust also accumulates more on the Martian equator than at the poles; NASA employees with a trained eye can look at a meteorite crater on Mars and tell the approximate latitude of where the impact happened just from the way dust is blown out – and many old craters show "equatorial" craters far from the current equator on Mars. Previous pole locations before movements of roughly 90 degrees include Arcadia Planitia and Utopia Planitia.

"Curved features on Jupiter's moon Europa may indicate that its poles have wandered by almost 90 degrees… the axis had shifted by approximately 80 degrees… its outer ice shell is decoupled from the core by a liquid layer."[85] In 2017, NASA said data from the Cassini probe suggest that Saturn's moon "Enceladus rolled onto its side" after something "triggered a redistribution of mass that destabilized the moon's rotation" and caused "a dramatic shift in the moon's polar axis."[86] With a mind closed to new perspectives, it is easy to overlook evidence of pole shifts. But once such evidence has our attention, and we appreciate it in the context of pole shifts, it is very difficult to dismiss. "The truth is out there" and the evidence will not be silenced.

The Great Flood – Was Not Caused by a Pole Shift

Gerald Massey wrote dozens of influential books on ancient history, mythology, astronomy, and religion in the late 19th century. When he commented on Egyptian and Assyrian mythology and said "The deluge here was evidently the result of a change in the pole-stars"[87] his comments were ahead of his time. [They may also have been referring to a far earlier event.] But his comments helped to provide fertile ground for the idea that the Great Flood might have been caused by a pole shift. By the middle of the 20th century the concept of pole shifts had become widespread and many people who knew what pole shifts are commonly thought of Noah's Flood as an ancient example of one.

There is a Book of Noah that did not make it into the Bible we know today. Despite being referenced in Jubilees, and the Book of Enoch, and despite being referred to in the Dead Sea Scrolls – there are too many missing parts and inconsistencies to know for sure what was originally written. There are various sources alleging contents of the Book of Noah that are not agreed upon. But with pole shifts in mind, Noah 65:1 seems especially relevant: "Noah saw the Earth had tilted and that its destruction was near." We just can't know for certain if such a verse definitely existed.

For many years during my early research on pole shifts, I assumed that Noah's Flood was global in extent and was caused by a pole shift. But far too much evidence suggests otherwise, and I had to eventually give up these long held assumptions. As I researched many subjects I also came to realize that I could somewhat precisely date the last pole shift, but I could not do so for Noah's Flood. The last pole shift made relatively permanent changes to the planet and there is evidence throughout many scientific fields pointing to a single time frame for it. The effects of the Great Flood alone were regional and temporary, on the scale of about a year or two – depending on your criteria for normalcy – before nature went back to normal. (A tsunami would be done in a day, but washing away sea salt, letting flooded areas dry, and allowing plants and animals and people to reestablish themselves in devastated areas would take longer.) As William Hutton and Jonathan Eagle wrote in <u>Earth's Catastrophic Past and Future</u>: "Note that Noah's Flood that buried inhabited land was temporary – roughly a year and a half according to Scripture."[88] The last "Great Flood" stories seem to describe not a pole shift but a different event without much clear evidence of a long-term impact on the planet.

The Bible's story of Noah, and other similar stories like Babylonia's Epic of Gilgamesh or the Greek story of Deucalion are the best evidence the Great Flood happened at all. Unfortunately, using the Bible to come up with historically accurate dates only works back to approximately 500 B.C. The Flood came long before that, and both the Bible and historical records combined prove increasingly vague the farther back in time we look.

Even the destruction of the Jewish Temple by the Babylonians, commonly dated to 586 B.C. and backed up by detailed Babylonian records, is incorrect. In a previous book I detailed how this incorrect date was reached by historians who did not take into account the differences between how Jews and Babylonians recorded years of kingship or marked the beginning of the year at different times. If a Babylonian had told us that the temple in Jerusalem was destroyed in the 19th year of Nebuchadnezzar's reign, he would have meant 586 B.C. But a Jew named Jeremiah told us, and he did not use the Babylonian reckoning of kingship (backdating to the start of the Babylonian year in the spring) he used the Jewish reckoning (postdating to the start of the new year in the fall.) Since Nebuchadnezzar took the throne in the second month of the Jewish calendar (in the fall) there is an 18 month difference in when the two peoples counted the first year of his reign. Long story short, the Temple was clearly destroyed in 584 B.C., but few historians realize this. (For a detailed explanation, read <u>End Times and 2019</u>, pp. 38-44)

For Noah's Flood, we must go much farther back in time, and the sources and dates get even sketchier the farther back we go. King Solomon may have started building the temple in 967 B.C. His reign is generally believed to have started in 970 B.C. and 1 Kings 6:1 says he started construction in his fourth year as king. But the Book of Kings gives lengths of reigns for all the kings of Israel (and Judah) and if we add those we would conclude that the temple stood for 429 years – 430 when we consider that the last king was deposed a year before the temple was destroyed – leading us to a date of 1014 B.C. for the construction of the temple. The Bible offers several other methods of calculating the start of construction, but they all contradict each other. All we know is an approximate date of roughly 1000 B.C.

We could try to add 480 years between the temple and the Exodus from Egypt, and use the genealogies listing the lifespans and age at fatherhood for all the patriarchs back to Noah. If we follow the numbers we might conclude that the Flood was in 2304 B.C. Or 2449 B.C. James Ussher suggests 2349 B.C. Most dates calculated from a biblical chronology lead to a Flood date between 2300-2400 B.C. If we accept that the biblical chronology used to calculate the date of Noah's Flood leads us to approximately 4,350 years ago – if this estimate is even close to reality – then it is not close to the date when the last pole shift apparently took place, and we need to consider a different cause for the Great Flood.

The last pole shift seems to have a lot of evidence supporting a specific time frame about 12,900 years ago. The Younger Dryas (a sudden cooling of some regions dated to 12,900+/-100 years ago) the Pleistocene Extinction (12,900 years ago) the postglacial return of flora and fauna to eastern Canada (12,240-12,620 years ago[89] the end of the Ice Age in North America (11,700 years ago) the sudden rise in North Atlantic water temperature (11,700 years ago) after the former ice caps finished melting into the oceans, the start of the warmer Holocene Age (11,700 years ago) once the warmer ocean allowed the Gulf Stream to warm up Western Europe, and Plato's dating of the final demise of Atlantis (11,600 years ago) all of these (and other evidence) point to a cataclysm somewhere around 12,900 years ago – three times as far back as the story of Noah – a miss of about 8,000 years.

I think I can make sense of all the dates above except for Noah's Flood. I know the Maya were obsessed with the years when the winter solstice sun would be in closest alignment with the galactic center and believed that the end of the present world would occur when the winter solstice sun completely passes the galactic equator. Enough so that Maya scholar John Major Jenkins asked: "is there some scientific basis for the idea that when the winter solstice sunrise 'stands' on the Galactic Center that any unusual physical effects might be expected? Today, science answers in the negative."[90] Due to the cycle of precession of the equinoxes – a slow and circular drifting of the earth's axis, drawing a circle in the sky over approximately 25,920 years – we know this alignment with at least part of the sun's disk lasts from around 1980-2018, with 2019 being the first year the winter solstice sun is completely past the galactic equator.

Ancient texts associate the summer solstice sun's alignment with the galactic center with a similar catastrophe, which would have been one half precession cycle back, about 12,960 years ago. If a pole shift occurred then, it would have moved the surface of the earth over the core, and Hudson Bay, at the North Pole until that most recent pole shift, moved thirty degrees south. This sudden move would cause massive earthquakes, tsunamis, and volcanic eruptions, and would explain the Pleistocene Extinction and the sudden cooling in some regions during the Younger Dryas. The North American ice cap, which caused the "ice age" in North America (much like Greenland or Antarctica have an "ice age" now) would then have started melting at its new ranger of warmer latitudes. It may have taken over a thousand years for the former ice caps to entirely melt and for sea levels and temperatures to stabilize.

Millions of cubic kilometers of glacial ice would have melted away in a short amount of time, quickly lowering the average water temperature in the North Atlantic, and keeping the temperature low until there was nothing left of the former ice cap to melt. If this took about a thousand years, it would explain the sudden rise in ocean temperature and the new warmer Holocene Age that started around 11,700 years ago. If there was a civilization like Plato described when the pole shift occurred, and it wasn't completely destroyed by the earthquakes, tsunamis, and climate change of the initial pole shift, its remaining coasts and cities may have been submerged by rising ocean levels, since the old ice cap melted much faster than the new one accumulated snowfall. I think that reasonably explains the dates for the pole shift evidence cited above.

But that doesn't lead us into the time frame of Noah. The Great Flood seems to be a much later event. Noah's story also lacks many characteristics of "mythological" stories that seem to describe the last pole shift. There is no mention of earthquakes, volcanoes, the sun veering off course, or the sky falling. There is no mention of the heavenly bodies not following their proper paths. No mention of a change in climate before and after the flood. No hint of a reference to the galactic center or any of the zodiac constellations. The Book of Genesis only mentions the fountains of the deep rising up, and rain falling down. Our only specific clue on timing, which seems to rule out being coincident with a pole shift around the summer solstice (June 22) or the winter solstice (December 21) is that Noah's Flood began on the 17th day of the month of Cheshvan, around November 1st in our calendar.

I soon realized that something big happened around 5,100 years ago, and history seems to have begun anew. Many cultures created a calendar that seems like an attempt to back-calculate to this event. India's Kali Yuga Age is believed to have begun in 3102 B.C., (sometimes dated to February 17th, much like Noah's Flood dates are all on the 17th of the month) the Mayan Long Count Calendar began in 3113 B.C., and Egyptian history begins with the unification of Upper and Lower Egypt into one kingdom around 3100 B.C. As I noted in <u>End Times and 2019</u>: "In 1923 Leonard Wooley led a dig in Iraq below the remains of the tower of Ur. At a level believed to be over 5,000 years old, they hit a layer of clay over ten feet thick. Below the sedimentary clay were the potsherds and flint tools of a very primitive society. Wooley telegrammed: 'WE HAVE FOUND THE FLOOD.'"[91]

As Bruce Scofield wrote in "What Really Happened in 3100 B.C.": There are only a few turning points (or watershed periods) in history that have had significant widespread effects... 3100 B.C. (+/-100 years) is the most enigmatic... It's possible that this period marks the most decisive time in the entire history of civilization."[92] He noted the abrupt appearance of civilization in Sumeria and Egypt, ...and a host of other areas where civilization seemed to suddenly appear without previous traces, as if everything before that had been washed away...

I also like astronomical clues, and astrology starting the circular zodiac somewhat arbitrarily with the Age of Taurus may indicate that civilization started in that age, which lasted from approximately 4,300-2,100 B.C. I realize that modern astrologers begin with the Age of Aires and this is the starting zodiac sign for the Ptolemaic system today, but ancient astronomers formed the zodiac with a start in the Age of Taurus, when the spring equinox sun would have been in the constellation. It may also be why Hathor, Egypt's goddess of destruction, is depicted wearing a "hat" made of the sun disk in the horns of a bull – if the last perceived episode of her wrath against humanity occurred with the vernal equinox sun in Taurus. The glyph for Taurus looks very much like our Phoenician letter "A," the first Greek letter alpha denoting the beginning, and the first Hebrew letter aleph, which even translates as ox or bull. These beginning letters shaped like the bull head and horns may also indicate a new beginning in Taurus.

Genesis 7:11 tells us that the Great Flood began on the seventeenth day of the second Hebrew month of Cheshvan, which typically would be in late October or early November. We cannot be certain of the exact date of Noah's Flood on our solar calendar in an unknown year, but this could easily have been around November 1, and the year was probably shortly before 3,100 B.C. That means the date was nowhere near the winter or summer solstice – times when a solar alignment with the galactic center seems to correspond to galactic superwave events and pole shifts. But the timing does correspond with the Taurid meteor shower – an annual concentration of meteors that appear to come from the direction of the constellation Taurus.

The trail of the disintegrated comet Encke loops around the sun and crosses Earth's orbit, with pieces of its debris falling to earth and in the Taurid meteor shower every fall. The looping orbit of the disintegrated comet debris actually crosses Earth's orbit twice a year, with peak meteors showers around November 1, and are visible again to a lesser extent around June 30, when the meteors appear to come from the direction of the sun and we cannot see them. But we should not ignore the significance of the second June 30 peak – it was June 30, 1908 when something crashed into Tunguska, Siberia – releasing approximately 10-15 megatons of explosive energy and knocking down two thousand square kilometers of trees. If the ocean had been hit that day instead, Earth's human population would probably be lower today, and we might have another much more recent flood story.

A most interesting star cluster, the Pleiades, is visible to the naked eye in the constellation Taurus – about where the bull's shoulder would be. And Hebrew tradition tells us that stars fell from the Pleiades at the time of Noah's Flood. The Talmud says: "When the Holy One... wanted to bring a flood upon the world, He took two stars from the Pleiades and brought a flood upon the world." I believe we can safely assume that our ancestors witnessed two large and bright meteors in the sky just before they were hit by the tsunami of the Great Flood that resulted from at least one of the meteors impacting the Indian Ocean.

We can also see this Taurid meteor event depicted in ancient depictions of Mithras, a Middle Eastern deity whose popularity faded out as Christianity spread. Mithras is usually shown plunging a sword into the shoulder of a bull (into the Pleiades of Taurus) with a cluster of seven stars near his head (the Pleiades) with blood gushing (meteors falling) out of the bull's shoulder. Left and right of this central act, cherubim are shown holding a torch, sometimes with both torches held upside down, aimed down towards the earth, sometime with one aimed up and one aimed down. At least one "flaming torch" seems to have descended to Earth from the direction of Taurus.

Most modern catastrophe scholars who study the Holocene Impact Working Group's analysis of Burckle Crater believe that Noah's flood was caused by a comet or large meteor crashing into the Indian Ocean at the impact site discovered in 2006 off the coast of Madagascar. Whatever crashed there went through over two miles of ocean and still made a crater 16-18 miles across on the seafloor. It is assumed to have been a meteor over a mile wide, moving at over twenty miles per second.

Using a wonderful online "Asteroid Impact Crater Calculator"[93] I entered an estimated diameter of 2000 meters, a density between that of iron and nickel (which most meteorites are made of) at 8 grams per cubic centimeter, a velocity of 32 kilometers per second, and a graze angle thirty degrees off from horizontal. The resulting estimate of the force of the impact of this object when it hit the Indian Ocean: approximately 4,085,068 megatons of TNT. That seems really high, so I thought more about my input data, and entered new numbers at the lowest possible end of every range. I lowered the diameter to 1800 meters, (less than the estimated size of the Burckle Impact Object) I lowered the density to 4 grams/cc, (less than the density of major asteroids like Pallas or Vesta) velocity to 30 kps, and graze angle to 45 degrees. Even that still leads to the result that the impact was equivalent to of 1,539,156 megatons of TNT.

For comparison, one commission in 2009 concluded that all the nuclear weapons of all the nuclear powers combined were recently estimated to have a combined yield of approximately 6400 megatons. That figure sounds low to me, and more have been made since then, so let's double that figure to 12,800 MT. Even when we double the nuclear weapons yield and artificially downgrade the likely characteristics of the Burckle Impact Object, then we would still have to imagine every nuclear weapon ever made exploding in the same spot simultaneously, over 100 times for each nuclear weapon, to equal the force of this impact into the ocean. Or we could say that compared to the Tunguska blast that knocked down 2000 square kilometers of trees, this impact in the Indian Ocean was probably over 100,000 times more intense. Even if my results are still too high by a factor of ten – that still makes for one heck of a splash. Yet it is still nowhere near as energetic or as devastating as the global tsunamis during a pole shift.

We can only speculate as to the height of the tsunamis that would have ravaged East Africa, the Middle East, South Asia, Indonesia, and Western Australia… But the common estimate is that they would have averaged about 600 feet high when they made landfall – even higher in Madagascar, which is the closest land to the impact crater. The northern part of the tsunami would have been funneled and condensed by the coastlines of the Arabian Sea and the Persian Gulf, and may have travelled an estimated 500 miles inland up the Tigris-Euphrates Valleys in Iraq, where civilization existed and the devastation may have been most intense. Did a tsunami push Noah's ark northwest up the river valley to Mount Ararat?

Egyptian civilization would have been better protected, as a tsunami from the Indian Ocean would have to flow over or around the Horn of Africa where Ethiopia and Somalia would have borne the brunt of it before it the waves could surge up the Red Sea to the southeast coast of Egypt. Keeping in mind that Lower Egypt is the delta region down at sea level on the Mediterranean in the north, and that Upper Egypt is the region of higher elevation in the southeast (Lake Nasser, for example, by the Aswan Dam, is 600 feet above sea level) – it is interesting that Egyptian texts suggest the Great Flood came from upper Egypt and the south. In the 175th chapter of the <u>Book of the Dead</u>, in the <u>Papyrus of Ani</u>, the Flood is said to have begun in Upper Egypt before affecting all of Egypt. If the Flood had been from a pole shift, we would expect the Mediterranean coast of Lower Egypt to have been submerged first. It doesn't seem like Lower Egypt experienced the tsunami at all.

The Burckle Crater Impact event would kill approximately two billion people if it happened today. Such an event could have destroyed a huge region, and prompted descriptions of the entire world being flooded, even if it only devastated the portion of the world known to those describing the event. This would certainly justify the comments in Genesis 7:11 about "the day all the fountains of the great deep burst open, and the floodgates of the sky were opened." Water blown into the sky by the impact, high humidity at ground level from evaporation off of flooded terrain, timed to match the normal onset of the rainy season in November in the area of Iraq make it very likely that Noah's story of a long period of non-stop rain could have occurred. Rain was incidental to the Flood; a side-effect that added to it – but rain was not the main cause.

Did a mass extinction impact from the Taurids occur around November 1 about 5,000 years ago, and lead to the formation of holidays commemorating the mass deaths from the event? "The festival of the dead amongst the ancient Peruvians was [already] celebrated on the same day as by the Spaniards, viz., on All Souls Day, November 2nd… the festival was generally observed in November south as well as north of the equator…" [which meant it was not merely about crops dying in the fall, as it would be spring south of the equator] "…how was this uniformity in the time of observance preserved, not only in far distant quarters of the globe, but also through the vast lapse of time since… the Indo-European first inherited this primeval festival from a common source?"[94] We should bear in mind that the tsunami would have passed South Africa and Australia, and would have still done significant damage on coastlines as far away as the Americas, even if the tsunamis there were a tenth or a twentieth as high as those in the Indian Ocean.

Halloween and the Celtic Samhain are on October 31, Catholicism's All Soul's Day is on November 2, Mexico's Dias de los Muertos (Days of the Dead) are on November 1 and 2, Haiti's Voodoo Fete Guede is on November 2. Ancient Egyptians remembered the slaying of the god Osiris, Lord of the Dead, on the 17th of the lunar month of Athyr or Athor (the same day Noah's Flood began in the Hebrew calendar, and a month linguistically equivalent to the goddess of destruction, Hathor.)

Charles Piazzi Smyth, in his influential 1865 book: <u>Life and Work at the Great Pyramid</u>, notes that the Hebrew and Chaldean name for the Pleiades star cluster in Taurus is Athor-aye. He also noted that global festivals of the dead often span three days and begin at sunset with the night of October 31. But most interesting of all is Smyth's comment that in Mexico: "at midnight as that constellation [the Pleiades] approached the zenith, a human victim… was offered up to avert the dread calamity which they believed impended over the human race… They had a tradition that at that time the world had been previously destroyed, and they dreaded lest a similar catastrophe would, at the end of a cycle, annihilate the human race."[95] The Mayan Lord of the Dead, Tzontemoc, was believed to have fallen to earth in early November. "Before the arrival of the Spaniards in Mexico, over 400 years ago, and probably much earlier, the Mexicans told of certain stars called Tzontemocque or Falling Hairs, which fell from heaven to earth with the Lord of the Dead. Their fall was commemorated annually in the Quecholli festival, said to have been held towards the end of October. This festival, and the falling of the stars, was associated with the end of the world."[96]

In Polynesia the dead were honored with a Feast of the Ancestors in early November. India has a Festival of Durga for the dead in early November, a month with the same name as the Pleiades. It could all be a coincidence. Or the vernal equinox sun could have been in Taurus (making it the Age of Taurus) when "Hathor" devastated Egypt and many other regions around the Indian Ocean in the form of a comet or meteor impact from the Taurid meteor shower, so named because they appear to come from the constellation Taurus.

After reviewing the available clues/evidence, I think the cause of Noah's Flood has probably been found by the Holocene Impact Working Group. The Burckle Crater site seabed sediments show concentrations of metals and other chemical compounds that must have precipitated out of boiling hot water. Drill cores include melted rock, glassy tektites from melted sand, and high levels of iron, nickel, and other metals that make up the majority of most meteorites. There are huge chevrons in Madagascar – wedge shaped deposits twice the size of Manhattan, and hundreds of feet high - that contain both deep ocean fossils and the same concentrations of metals commonly found in meteors. Similar, but smaller chevrons in Western Australia and other more distant coasts allowed scientists to triangulate the position of the crater. One autumn day about 5,000 years ago, a huge meteor fell into the Indian Ocean and reduced existing civilizations near its coasts to a few survivors in the mountains. Despite the near certainty many of us once had on the subject – the Great Flood wasn't caused by a pole shift. Despite the catastrophic destruction of this very real flood event, we must look even further back in time to understand the cycle of pole shifts.

Ancient Comments on Pole Shifts

Ipuwer

Some Egyptian records may describe a previous pole shift. The Ipuwer Papyrus, a document written by a sage named Ipuwer around 4,000 years ago, includes the descriptions: "the earth turned upside down," "the land turns round like a potter's wheel," "the land is not light," and "I show thee the land upside down; it happened that which never had happened." We can't be certain of the time frame to which these descriptions apply. They could be literal references to an ancient pole shift, but they might merely have been social commentary on current events, worded in a way that happens to draw our attention to pole shifts. Perhaps the writer was being poetic, using metaphor and simile, just as someone in modern America might say "the nation is being ripped apart" by partisan politics. The descriptions in the Ipuwer Papyrus could merely refer to social unrest. It has been suggested that this incomplete document probably ended with a prayer "to the Lord of All" and "prophesying the coming of a powerful king who would restore order." It is easy to see why at least some interpreters suggest this papyrus is describing a cataclysm, the hope of some kind of messianic redemption, and a pole shift.

Hesiod

In approximately 700 B.C. Hesiod gathered various Greek creation stories and wrote the Theogony – the genealogy of the gods. It detailed the Titanomachy – "The war of the gods" in which the Olympian gods led by Zeus fought and overthrew the Titans led by Kronos (who had previously overthrown his father, Uranus... In every changing of the divine guard, mankind is destroyed and re-created, and the gods who lost were cast down to Tartarus, much like stars being cast down below the horizon when the poles shift and certain constellations will not be seen again. (It sounds a lot like Satan and his angels – quite possibly represented by the polar constellation Draco falling from its polar zenith in the sky, and stars that fall below the observer's horizon – that will be cast down out of heaven in Revelation 12:9 – "The dragon and his angels waged war... and there was no longer a place found for them in heaven.") In the Greek stories, sometimes such cast down gods/stars are freed and brought back up thousands of years later, only to be eventually cast down and imprisoned in Tartarus again. In the Book of Revelation, Satan is cast down, and "bound for a thousand years...in the Abyss... After that, he must be set free." Does Revelation basically describe the same banishment of defeated gods as Greek mythology?

In all, Hesiod detailed five ages of man – a Golden Age of peace and harmony in which the gods were present with man, followed by declining ages in which the gods left Earth for heaven: the Silver Age and the Bronze Age. The Heroic Age was a slight improvement, and then the final Iron Age was worse. We are living in the fifth (Iron) age now, an evil, miserable age in which harmony with the divine and the Earth and greatness and goodness have evaporated away. It was believed that Zeus himself will bring an end to this age, as he has done before.

In Hesiod's time, knowledge and technology and prosperity were increasing. Greek civilization was spreading. We might expect the Greeks would have had an optimistic view of the future based on recent progress; much like Western society has today. But Hesiod and the rest of Greek culture believed that the Golden Age was in the distant past, and that we are due for destruction again at the end of the present age.

Heraclitus

Heraclitus was a Greek philosopher who wrote intentionally obscure commentary "On Nature" in the years around 500 B.C. Cryptic, paradoxical, and full of riddles, his thoughts – mostly on the nature of the universe, of the divine, the constant changes and evolution and becoming of things, and on humanity's general lack of understanding – were difficult to understand. He was known as "the obscure" and "the riddler." What he wrote is only known to us by fragments quoted by other writers like Diogenes. But we must assume that despite being intentionally difficult to understand, Heraclitus was widely respected and quoted often.

It is widely believed that Heraclitus wrote about a "fated necessity" – a cosmic cycle of destruction through periodic conflagration. Theophrastus seems to say that Heraclitus had a definite time frame of 10,800 years between such events, but evidence for this specific duration of his cosmic great year is indirect. Others say he felt 18,000 years was the duration of the cycle. If Noah's Flood had spilled over the Middle East from the Indian Ocean (I don't think it did) and affected Greece a mere 2,500 years before Heraclitus, we would expect some mention of the event – or at least, belief in a shorter cycle and more recent end of a cycle event. Instead, Heraclitus' time frame roughly corresponds to half a cycle of precession.

Herodotus

The Greek historian Herodotus was one of many respected Greek wise men who may have been initiated into an ancient Egyptian brotherhood which knew the details of cycles of cataclysms but felt that while the knowledge must be preserved – it should only be known to the brotherhood, and not publicized to the world. "Herodotus often mentions the obligation he was under to remain silent concerning 'sacred' subjects."[97] His feelings on this come from a widespread ancient tradition. "Egyptian priests were bound by oath to convey their knowledge to outsiders only indirectly through fable and parable."[98] Many secretive groups still hold fast to this idea today.

Writing around the year 446 B.C., Herodotus did record some details, including that "...on the accounts given me by the Egyptians and their priests. They declare that three hundred and forty-one generations separate the first king of Egypt from the last I have mentioned – the priest of Hephaestus – and that there was a king and a high priest corresponding to each generation. Now to reckon three generations as a hundred years, three hundred generations make ten thousand years, and the remaining forty-one generations make 1340 years more; thus one gets a total of 11,340 years, during the whole of which time, they say, no god ever assumed mortal form; nothing of the sort occurred either under the former or under the later kings. They did say, however that four times within this period the sun changed his usual position, twice rising where he normally sets, and twice setting where he normally rises."

Four pole shifts in less than 12,000 years would be more often than any other source indicates... and I have other reasons to believe that the Shemsu Hor, "The Followers of Horus" – the brotherhood of Egyptian wise men that kept ancient knowledge intact over many thousands of years – actually have a historical record (a King's List) of almost 40,000 years... and that "this period" in which four reversals or pole shifts occur is over the larger period of the Egyptian King's List...

Herodotus elsewhere describes different groupings of Egyptian gods coming into existence in different ages: "even Dionysos, the youngest of the three, appeared, they say, 15,000 years before Amasis. They claim to be quite certain of these dates."

If the same four pole shifts are allowed to occur over that larger time scale, then the records may match several other sources... Four previous worlds are mentioned in the creation stories of the Maya, the Nahuas, the Pueblos, the Navahos, Hindus, Tibetan Buddhists, Persian followers of Zoroaster, in Iceland, in China, in Polynesia... Herodotus' rough calculation that some great change in Egypt led to the disappearance of previous gods and the reign of mortal kings roughly 11,800 B.C. still gets my attention.

Plato

The Greek philosopher Plato was a central figure among the Greek philosophers and the founder of the Academy in Athens. His teacher Socrates, his indirect mentor Pythagoras, and Plato's own student Aristotle form the pinnacle of ancient Greek thought. Plato wrote many books between approximately 400-350 B.C., including Critias, Dialogues, Laws, The Republic, Timaeus, and many others. In these books he repeatedly referenced an intellectual heritage from the very distant past. Plato said that Egypt had experienced a very long period of stability and was the main source of ancient wisdom being preserved into his own time. On the stability and duration of Egyptian civilization, Plato commented: "you will find that the works of ten thousand years ago – I mean that literally – are no better or worse than those of today."[99]

Plato said that Egyptian priests knew of ancient cycles of destruction which he learned about through Mystery schools. Unfortunately he was obliged to discuss such knowledge indirectly. Initiates in such schools are bound by honor and oath not to openly discuss great teachings with those who lack the background to fully understand, lest important truths end up distorted and lost. Plato said: "This is the reason why every serious man in dealing with serious subjects carefully avoids writing about them, lest he may thereby cast them as a prey to the envy and stupidity of the public"[100] as pearls before swine.

Of all the Greek thinkers, Plato was the one most in tune with ancient wisdom, and he understood the unique language of myth. As the authors of Hamlet's Mill point out early in their book, most ancient myths explain "the theory about 'how the world began' [which] seems to involve the breaking asunder of harmony, a kind of cosmogonic 'original sin' whereby the circle of the ecliptic (with the zodiac) was tilted up at an angle with respect to the equator."[101] Plato's understanding of this theme is obvious in his writing, so it is no surprise that in Hamlet's Mill they also note that "Plato did not escape the idea he had inherited, of catastrophes and the periodic rebuilding of the world."[102] Plato described catastrophes that repeatedly brought world ages to their ends, and he is our primary source of information on Atlantis and its destruction. But when Plato wrote about Atlantis, we should understand that he masked certain details from the uninitiated.

Like a "living Rosetta Stone," Plato translated the extremely ancient initiated wisdom he learned into more modern stories which still held the keys of the brotherhood's knowledge but conveyed messages on both obvious and hidden levels... "Plato did not invent his myths, he used them in the right context... without divulging their precise meaning: whoever was entitled to the knowledge of the proper terminology would understand." There was a "strict secrecy that surrounded archaic science"[103] and those initiated into the ancient brotherhood were bound not to explain everything too clearly to the uninitiated masses.

As John Michell wrote in The Dimensions of Paradise, Plato: "hints at knowledge of the code of numbers behind the structure of the universe, but it seems as if he was inhibited from imparting its details... Plato studied sacred science with the priests in Egypt... and he also went to Phoenicia to acquire the science of the Persian magi. At home he was initiated into the Pythagorean system... the knowledge was not his own but had been entrusted to him under an oath of secrecy."[104]

Plato and Pythagoras believed "the tenet of traditional philosophy that number [in the mind of God] preceded creation and determined its development."[105] In Book 3 of Plato's <u>Republic</u>, Socrates, speaking for the gods, suggests that the citizens of the ideal city should gather in separate groups representing gold, silver, bronze, and iron. (This indicates four ages...) By marrying at a certain point in a great cosmic cycle, as indicated by the nuptial number, the guardians' offspring are most likely to be golden. Taken literally, this idea of marrying under astronomical guidance reeks of eugenics and astrology and numerology. But I think the premise gives clues to the astronomically determined heavenly wedding, the marriage of heaven and earth that periodically takes place during a pole shift event, with a huge influx of cosmic energy inseminating new life on a new earth.

Plato also believed there was a recurring and periodic marriage of heaven and earth linked to alternating catastrophes through fire and water. His calculation of the nuptial number that determined this periodicity is complex and there is no one consensus on it. 12,960,000 years[106] is one choice, suggested in Plato's Republic as 3^4 times 4^4 times 5^4 but the length of precession cycle (5 times 4^3 times 3^4) is also considered and mine is the more commonly assumed 12,960 years, or half a precession cycle. Aristotle may also have said 12,954 years; Tacitus said 12,854 and Cicero said 12,954.[107] While there may not be agreement on the duration of a world age in Plato's view, he did make a comment on: "the world, the formation of which is controlled by a large number."[108] Given Plato's love of easily divisible numbers, I think the precessional half-cycle of 12,960 years is the clear winner for his marriage number that gives birth to a pole shift and a re-created world.

Almost every ancient tradition, from Egypt to Greece to China to modern Masonic lodges – is very interested in the relationship between the circle and the square as a representation of heaven and earth. The Great Pyramid in Egypt has a base perimeter approximately 6.28 times its height – the same 2π ratio of the perimeter of a circle to its radius. Each side of the base is 365.25 royal cubits across, the same as the days in one orbit of the Earth around the Sun. The pyramid represents the Earth – or at a minimum, the above ground portion represents the Northern Hemisphere at a ratio of 43,200 to 1. In a sense, the builders succeeded in squaring the circle, my incorporating certain

Earth measurements and geometric ratios into its construction. So it may be important that the height of the missing capstone, were it in place to be measured, could tell us when the Earth reaches completion, and a new cycle begins. Some say the height would be 5780 inches, and that Hebrew Calendar year 5780 (2019-2020 A.D.) could be the year to watch for.

Peter Lemesurier, in <u>The Great Pyramid Decoded</u>, also noted that the distance from one corner of the pyramid to the opposite (the diagonal of its base) – the distance one needs to cross to end up half way around the pyramid at the opposite corner, is approximately 12,828 inches.[109] This is remarkably close to estimates for "crossing" one half precession cycle in years, and may warn of the time between marriages of heaven and earth – the time between pole shifts.

In "church architecture that the central dome represents Heaven while the square shape of the nave represents Earth… the relation of the circle to the square is found quite universally from China to India and Persia, from Rome to Christianity and Islam."[110] Solving the riddle of "squaring the circle" pertains to the marriage of heaven and earth.[111]

Steven Sora seems to agree: "Plato's V Book of <u>Laws</u> said the number 12,960 was the sacred marriage of the circle and the square"[112] and we know the circle and the square indicates heaven and earth. "Plato… [held] the traditional belief that civilization is periodically destroyed by cataclysms due to some natural, or divinely willed event."[113] Does something amazing happen that unites heaven and earth and changes them roughly every 12,960 years?

As for Atlantis, Plato tells us that this great nation was destroyed approximately 9,600 B.C. This timing coincides with a major and sudden rise in the temperature of North Atlantic seawater, and apparently marks the end of over 1,000 years of melting glaciers at the former poles of the earth, after a pole shift had moved former ice caps to warmer latitudes.[114]

In Plato's <u>Timeaus</u>, and also in <u>Critias</u>, written around 360 B.C., we are told how the Egyptian high priest Sonchis told Solon, the grandfather of Socrates, about Atlantis:

"Tell us," said the other, "the whole story, and how and from whom Solon heard this veritable tradition."

He replied: "At the head of the Egyptian Delta, where the river Nile divides, there is a certain district which is called the district of Sais, and the great city of the district is also called Sais, and is the city from which Amasis the king was sprung. And the citizens have a deity who is their foundress: she is called in the Egyptian tongue Neith, which is asserted by them to be the same whom the Hellenes called Athena.

Now, the citizens of this city are great lovers of the Athenians, and say that they are in some way related to them. Thither came Solon, who was received by them with great honor; and he asked the priests, who were most skillful in such matters, about antiquity, and made the discovery that neither he nor any other Hellene knew anything worth mentioning about the times of old.

On one occasion, when he was drawing them on to speak of antiquity, he began to tell about the most ancient things in our part of the world--about Phoroneus, who is called 'the first,' and about Niobe; and, after the Deluge, to tell of the lives of Deucalion and Pyrrha; and he traced the genealogy of their descendants, and attempted to reckon how many years old were the events of which he was speaking, and to give the dates.

Thereupon, one of the priests, who was of very great age; said, 'O Solon, Solon, you Hellenes are but children, and there is never an old man who is a Greek.'

Solon, hearing this, said, 'What do you mean?'

'I mean to say,' he replied, 'that in mind you are all young; there is no old opinion handed down among you by ancient tradition, nor any science which is hoary with age. And I will tell you the reason of this: there have been, and there will be again, many destructions of mankind arising out of many causes.

There is a story which even you have preserved, that once upon a time Phaethon, the son of Helios, having yoked the steeds in his father's chariot, because he was not able to drive them in the path of his father, burnt up all that was upon the earth, and was himself destroyed by a thunderbolt. Now, this has the form of a myth, but really signifies a declination of the bodies moving around the earth and in the heavens, and a great conflagration of things upon the earth recurring at long intervals of time: when this happens, those who live upon the mountains and in dry and lofty places are more liable to destruction than those who dwell by rivers or on the sea-shore; and from this calamity the Nile, who is our never-failing savior, saves and delivers us.

When, on the other hand, the gods purge the earth with a deluge of water, among you herdsmen and shepherds on the mountains are the survivors, whereas those of you who live in cities are carried by the rivers into the sea; but in this country neither at that time nor at any other does the water come from above on the fields, having always a tendency to come up from below, for which reason the things preserved here are said to be the oldest.

The fact is, that wherever the extremity of winter frost or of summer sun does not prevent, the human race is always increasing at times, and at other times diminishing in numbers. And whatever happened either in your country or in ours, or in any other region of which we are informed--if any action which is noble or great, or in any other way remarkable has taken place, all that has been written down of old, and is preserved in our temples; whereas you and other nations are just being provided with letters and the other things which States require; and then, at the usual period, the stream from heaven descends like a pestilence, and leaves only those of you who are destitute of letters and education; and thus you have to begin all over again as children, and know nothing of what happened in ancient times, either among us or among yourselves.

As for those genealogies of yours which you have recounted to us, Solon, they are no better than the tales of children; for, in the first place, you remember one deluge only, whereas there were many of them; and, in the next place, you do not know that there dwelt in your land the fairest and noblest race of men which ever lived, of whom you and your whole city are but a seed or remnant. And this was unknown to you, because for many generations the survivors of that destruction kept no records..."

As Plato wrote in <u>Laws</u> over 2500 years ago:

"Athenian: Do you consider that there is any truth in the ancient tales?

Clinias: What tales?

Athenian: That the world of men has often been destroyed by floods, plagues, and many other things, in such a way that only a small portion of the human race has survived.

Clinias: Everyone would regard such accounts as perfectly credible.

Athenian: Come now, let us picture to ourselves one of the many catastrophes,—namely, that which occurred once upon a time through the Deluge.

Clinias: And what are we to imagine about it?

Athenian: That the men who then escaped destruction must have been mostly herdsmen of the hills, scanty embers of the human race preserved somewhere on the mountain-tops.

Clinias: Evidently.

Athenian: Moreover, men of this kind must necessarily have been unskilled in the arts generally, and especially in such contrivances as men use against one another in cities for purposes of greed and rivalry and all the other villainies which they devise one against another.

Clinias: It is certainly probable.

Athenian: Shall we assume that the cities situated in the plains and near the sea were totally destroyed at the time?

Clinias: Let us assume it."

The conversation implies it is common knowledge in pre-Christian Greece that past civilizations have been destroyed. In The Statesman, Plato adds that "At periods the universe has its present circular motion, and at other periods it revolves in the reverse direction… Of all the changes which take place in the heavens this reversal is the greatest and most complete… There is at that time great destruction of animals in general, and only a small part of the human race survives."

Plato had definitely learned about pole shifts.

Diodorus

The Greek historian Diodorus wrote the Bibliotheca Historica around 50 B.C., in which he details the ancient history of the world, including Egypt. Diodorus' account of Egyptian history said that there was a real "Hercules" (a.k.a. Herakles, Samson, and a bunch of other figures who have 12 labors indicative of 12 zodiac signs and known for knocking down two pillars symbolizing a pole shift) during the Age of Osiris ruling over Egypt. "The Egyptians reckon more than ten thousand years," Diodorus said, to the era when Herakles pushed down the pillars – more than 12,000 years ago by now.

After this the priestly astronomers, the Followers of Horus, ruled Egypt. Diodorus somehow specifically dates the catastrophic flood 10,628 years before his own time – and if we assume his birth was in 90 B.C., then this would be 12,736 years before 2018 A.D. – quite consistent with the idea of a catastrophe one half cycle of precession in our past. Interestingly, Diodorus also comments on the previous end of an even earlier world age, adding "'but as some write there are nearly 23,000' [years between the flood and his own life] as if this latter sum were based upon some other distinct and discordant reckoning… reckoned from a different and earlier commencement of mankind."[115]

This would place another destruction of the world over 25,000 years ago, or approximately one complete cycle of precession in our past… It matches the timetable for the end of Neanderthal Man (about 26,000 years ago) and evidence for a pole in the Greenland Sea until about 27,000 years ago[116] and the maximum extent of East Asian mountain glaciers about 26-27,000 years ago[117] and the first appearance of the North American ice cap in Ohio over 25,000 years ago.[118] Diodorus' "earlier commencement of mankind" almost 26,000 years ago seems to fit the timing two pole shifts back, when the North Pole arrived at Hudson Bay.

I am most impressed with the similarity of the dates given by Diodorus with my own conclusions that the last two pole shifts occurred one half and one full precession cycle in the past. He is not to be confused with Diodorus of Tarsus, who lived two centuries later; but this second Diodorus did make one comment worth mentioning here; he wrote about the Star of Bethlehem and suggested it was not a star but a "force that assumed the shape of a star" interpreted as a "periodic subtle energy outburst."[119]

Seneca

The Roman philosopher Seneca (the Younger) was a tutor and advisor to the Emperor Nero. One of his most famous plays, Thyestes, (written in 62 A.D.) was based on Greek Mythology of warring gods and kings – and is sometimes credited with starting the revenge tragedy genre. Hamlet and many other later plays fit this genre, but so do far more ancient stories with similar characters like Horus battling his evil uncle Set after the death of his father Osiris... More to the point, Seneca seems to have described an ancient pole shift in "Thyestes" when lamenting over the sun god Apollo's odd behavior:

"You've fled back and plunged the broken day out of the sky... And sent night from the east at a strange time to bury the foul horror in a new darkness... Sun, where have you gone? How could you get lost half way through the sky? ...The way things take turns in the world has stopped... I'm stuck with terror in case it's all collapsing... The zodiac's falling... Have we been chosen... to have the world smash and fall upon us?"

Seneca also wrote some commentary "discussing the Babylonian cosmology of Berossos... that the future flood will take place when [a certain] conjunction takes place in Capricorn. For the former is the constellation of the summer solstice, the latter of the winter solstice; they are the decisive signs."[120] I suspect the last pole shift started on the summer solstice, and that the next one will start at the winter solstice.

After Seneca

There seems to be a 1600 year gap in pole shift references after Seneca, because as Christianity spread, the Church formed and covered up evidence of anything that did not support their biblical narrative. The previously universal acceptance of the idea of "cyclical time suffered irreparable harm from the doctrine of the Incarnation."[121] "Gnostics were proud of the knowledge they had inherited from the ancient world… The Church, on the other hand, taught that the coming of Christ was a unique event… rendering all other religions obsolete."[122] Hinduism and many other cultures accept a cyclic view of history; Shiva and Vishnu come back at regular intervals to play the same recurring roles as destroyers and saviors in history… But Jesus saves us once; pole shifts and cycles of cosmic destruction do not agree with the once and done nature of linear progression in Christian teachings.

But even the Bible sometimes hints at cycles, such as in Revelation 22:13 when God says: "I am the Alpha and the Omega, the first and the last, the beginning and the end." Or in Ecclesiastes 1:9-11 "There is nothing new under the sun. Is there anything new of which one might say, 'See this, it is new?' Already it has existed for ages which were before us. There is no remembrance of earlier things; and of the later things which will occur, there will be no remembrance among those who will come later still." Even though I believe pole shifts are described at least a few times in Genesis and Revelation – in general Christianity describes man's existence in terms of linear progress. This way of thinking dominated Western culture after the first century A.D. and is probably the main reason we find no clear references to cyclic catastrophes or pole shifts in Western culture for about 1600 years after Seneca, until science began to gain ground over religious teachings.

Early Modern Writers on Pole Shifts

Since about 1960, space age technology has allowed research orders of magnitude beyond anything that came before. But even before we had sophisticated lasers and cameras and satellites to gather more and more evidence of pole shifts, earlier thinkers noticed the evidence right at our doorstep. With the end of the Middle Ages came an age of exploration and new scientific thinking that brought the idea of pole shifts back to the attention of learned men.

Was it because the Templars had discovered ancient wisdom (and ancient maps including the Americas) during the Crusades? Marco Polo was said to have brought a map of the entire world back from China, including "a great southern island continent." There is also the possibility that with the fall of Constantinople to the Turks in 1453, Byzantine scholars fled west with thousands of ancient maps and other documents from their libraries. With some large portion of all the ancient manuscripts that had been accumulated by Roman and Byzantine rulers, scavenged from Europe and the Middle East over many centuries suddenly making their way to Western Europe, it was only a matter of time before scholars started to put together some of the clues that revealed ancient knowledge.

Ancient maps were an important part of this accumulated wisdom. In an upcoming chapter we will review evidence that ancient mapmakers had charted the entire world, including Antarctica, using advanced mathematics like spherical trigonometry for their map projections – and that they had done so before Antarctica was covered in ice. For the moment it will suffice to suggest that within forty years of discovering certain maps from Byzantine archives, America was "discovered" by one Christopher Columbus, who some days before landfall, noted in his diary on October 4, 1492: "In the spheres which I have seen and in the Mappae Mundi (map of the world) it is in this region."[123] If Western European scholars suddenly gained access to a treasure trove of ancient wisdom including maps drawn both before and after the last pole shift, then they must have understood that the surface of the planet had undergone some massive changes.

That perspective would only have been reinforced by discoveries in the Americas. The traditions of Native Americans like the Aztecs, the Maya, the Inca, and the Hopi may have encouraged Europeans notice certain similarities in world mythology and reevaluate the possibility that there was historical truth in the old tales.

If anyone with power in Western Europe were convinced of this, they would have wanted to learn as much as possible about pole shifts. Masonic lodges began to spring up in England and France around 1720, and their members traced their origins to the Knights Templar, who had discovered some secret information under the Temple of Solomon during the Crusades. Not long after Masonic lodges began to dominate French and British intelligentsia in the late 1700s, the French army conquered Egypt in 1799, with a primary focus on bringing in "the Scientific Commission" of 167 scholars. There was little military reason for the scholars to be there, or for the French to be there at all. But their scientific mission in Egypt was considered very important. Too important for the British to leave them alone – so the British conquered Egypt from the French in 1801.

Both Britain and France began a long obsession with rediscovering ancient Egyptian wisdom. Egyptian obelisks were brought to London, Paris, and New York – where "Cleopatra's Needle" can be seen in Central Park. Was "Egyptomania" just a fad, or is there a lot more to it? Something got people to question uniformitarianism – the idea that everything has been stable and gradual processes have shaped the earth for millions of years. Something made people pay attention to the idea that despite our recent history of stability and gradual processes, there is evidence of repeated upheavals and catastrophes shaping the earth.

Entire books have been devoted to conspiracy theories involving Freemasons and the symbolism on the Great Seal of the United States, as shown on the back of the one dollar bill, with the all-seeing eye completing the unfinished pyramid. The Great Pyramid is known to represent the world, and the pyramid's capstone represents completion of the world. Does the eye in the sky represent the end of the present world? Is the timing of the present world's completion predictable, based on the measurements of the Great

Pyramid? One of the most famous pyramidologists, Charles Piazzi Smyth, said in his books that he believed that the Great Pyramid in Egypt "was an instrument of prophecy" whose measurements incorporated a message about a future end of the world.[124] It that the great secret of the secret societies?

Without going too far down that very interesting rabbit hole (many other books have already covered such topics) – it will have to suffice for the purposes of this book to say that altered global surface conditions and knowledge of pole shifts seem to have suddenly reappeared in European thinking after a very long absence, and there must have been some discoveries responsible for that.

Jakob Boehme

Boehme was an early 17th century German philosopher, theologian and mystic who believed he had special insights into the nature of the cosmos. Some consider him the greatest of Christian Gnostics, especially when it comes to his knowledge of myth and symbol regarding revelation and the end times. From 1612-1624 he wrote many scandalous books which did not conform to the religious views of his time, including <u>Aurora</u>, <u>Mysterium Magnum</u>, <u>Of the Earthly and of the Heavenly Mystery</u>, <u>Of the Last Times</u>, <u>The Signature of All Things</u>, and <u>Of the Last Judgment</u>. He made several prophetic predictions including a well-known one about Atlantis rising again, approximately in the early 21st century.

The American psychic Edgar Cayce was asked a question in 1932 starting with: "Three hundred years ago Jacob Boehme decreed Atlantis would rise again at this crisis time when we cross from this Piscean Era into the Aquarian..." Though Boehme did not specifically mention a "pole shift" – Edgar Cayce often did talk about expecting one near the turn of the millennium. The timing of our entrance into the Aquarian Age is not clear, but the French Institute for Geophysics uses the year 2010 as the approximate boundary between ages. I cannot see how a continent would rise up from under an ocean or ice cap again by any means other than a pole shift, so I include Boehme as acknowledging the idea of a pole shift in the early 21st century. Unfortunately I cannot find any details on any specific vision or calculation from which he formed this prophetic conclusion.

John Milton

John Milton's epic 1667 poem "Paradise Lost" may be the next clear reference (since Seneca) to anyone writing about pole shifts, as part of his story about mankind being kicked out of Eden says:

"Some say, he bid his Angels turn askance

The poles of earth, twice ten degrees and more,

From the sun's axle; they with labor pushed Oblique the centrick globe."

Milton clearly suggests in "some say" that earlier records suggest God and His angels turned the axis of the planet and moved the poles. I would love to know what ancient sources he had access to.

Thomas Burnet

In 1681, Thomas Burnet wrote in The Sacred Theory of the Earth that "The Poles of the world did once change their situation, and were at first in another posture from which they are now, till that inclination happen'd… earth changed its posture at the Deluge, and thereby made these seeming changes in the Heavens."

Isaac Newton

Sir Isaac Newton, the founder of physics, calculus, and laws of planetary motion, published Principia Mathematica in 1687. One comment in it is particularly interesting: "…let there be added anywhere between the pole and the equator a heap of new matter like a mountain, and by its perpetual endeavour to recede from the centre of its motion will disturb the motion of the globe, and cause its poles to wander about its surface."[125] Was this merely a deduction through scientific reasoning – or did he have evidence that it had happened before?

Newton believed that analyzing the numbers that kept popping up in both measurements of the cosmos, and in ancient monuments, and especially in clues in the Bible – would allow him to figure out anything and everything. John Maynard Keynes, the economist responsible for deficit spending and never ending inflation, held Newton's position at Cambridge a few centuries after Newton and had access to many of his private papers. In 1947 Keynes gave a speech on Newton and said:

"Newton was not the first of the Age of Reason. He was the last of the magicians, the last of the Babylonians and Sumerians... he looked on the whole universe and all that is in it as a riddle, as a secret which could be read by applying pure thought to certain evidence, certain mystic clues which god had laid about the world to allow a sort of philosopher's treasure hunt to the esoteric brotherhood. He believed that these clues were to be found partly in the evidence of the heavens... but also partly in certain papers and traditions handed down by the brethren in an unbroken chain back to the original cryptic revelation in Babylonia. He regarded the universe as a cryptogram set by the Almighty... By pure thought, by concentration of mind, the riddle, he believed, would be revealed to the initiate." I agree that Newton felt this way, and cannot emphasize enough that we should pay attention to these ideas – as Newton, the inventor of Calculus and Physics, was one of the most intelligent men who ever lived.

Georges Cuvier

One of the founders of paleontology, the French zoologist Georges Cuvier, was noted on the first page of this book as a champion of catastrophism: the theory that the Earth's surface features are best explained by a series of short-lived catastrophes that cause major changes in short periods of time. Cuvier published a groundbreaking report in 1813 titled: "Essay on the Theory of the Earth." It detailed the evidence for his radical and unpopular theories on fossils, the extinction of species, and natural catastrophes that periodically destroy most of the world:

"These repeated eruptions and retreats of the sea have neither been slow nor gradual; most of the catastrophes which have occasioned them have been sudden; and this is easily proved, especially with regard to the last of them, the traces of which are most conspicuous. In the northern regions it has left the carcasses of some large quadrupeds which the ice has arrested, and which are preserved even to the present day with their skin, their hair, and their flesh. If they had not been frozen as soon as killed they must quickly have been decomposed by putrefaction. But this eternal frost could not have taken possession of the regions which these animals inhabited except by the same cause that destroyed them; this cause, therefore, must have been as sudden as its effect. The breaking to pieces and overturning of the strata, which happened in former catastrophes, show plainly enough that they were sudden and violent like the last; and the heaps of debris and rounded pebbles which are found in various places among the solid strata, demonstrate the vast force of the motions excited in the mass of waters by these overturnings. Life, therefore, has often been disturbed on this earth by terrible events – *calamities which, at their commencement, have perhaps moved and overturned to a great depth the entire outside crust of the globe.*"[126]

At another point Cuvier wrote: "It is to fossils that we owe the discovery of the true theory of the earth; without them, we should not have dreamed, perhaps, that the globe was formed at successive epochs, and by a series of different operations. They alone, in short, tell us with certainty that the globe has not always had the same envelope..."

One of the anonymous book reviews in the London newspapers of early 1814 commenting on Cuvier's work confronted deeply held religious views and normalcy bias which made the book critic reluctant to accept Cuvier's numerous groundbreaking conclusions... but he was still able to summarize Cuvier's thoughts on the young age of the world's current surface conditions: "The original figure of the earth may have been extremely unlike the present; it may have been vastly irregular; and in the course of changes which it has undergone, the axis of rotation may have changed its position, and have passed through a series of variations..."

"...What has taken place on the surface of the globe since it has been laid dry for the last time, and its continents have assumed their present form, at least as such parts as are somewhat elevated above the level of the ocean, it may be clearly seen that this last revolution, and consequently the establishment of our existing societies, could not have been very ancient. This result is one of the best established, and least attended to in rational zoology; and it is so much more the valuable, as it connects natural and civil history together..."

Charles Darwin

The father of evolutionary theory, Charles Darwin, also recognized the evidence for the occasional destruction of the Earth's surface conditions, causing mass extinctions followed by the sudden spread of the most adaptable species to survive the cataclysms. In 1834, he wrote in <u>The Voyage of the Beagle</u>: "It is impossible to reflect on the changed state of the American continent without the deepest astonishment... To destroy animals, both large and small, in Southern Patagonia, in Brazil, on the Cordillera of Peru, in North America up to Behring's Straits, we must shake the entire framework of the globe. No lesser physical event could have brought about this wholesale destruction not only in the Americas but in the entire world."[127] In 1835, Darwin described an earthquake he experienced in Chile: "the world, the very emblem of all that is solid, moves beneath our feet like a crust over a fluid." He later added: "the earth's crust floats in a like manner on a sea of molten rock" and suggested that a large enough force could do this to "the entire globe."

Such catastrophes seem to be the primary impetus for evolution. Some populations move to colder climates or higher elevations with colder weather. Other populations move to lower altitudes and latitudes closer to the equator. The plants and animals either adapt or die, there are no other choices. Pole shift researchers like Charles Hapgood agree: "If we consider all the effects of crust displacement, both toward the equator and toward the poles, we can see that crust displacement constitutes the most powerful engine imaginable for forcing life forms to adapt to all possible habitats."[128] Most scientists will acknowledge a connection between rapid speciation and magnetic pole shifts but fail to make the connection to the crustal displacement that causes both. Geologist Joseph Meert recently suggested that "these steps in evolution are connected to rapid reversals in the direction of Earth's magnetic field."[129] This topic is also covered in depth in Robert Felix's book: <u>Magnetic Reversals and Evolutionary Leaps</u>.

I am glad that scientists make this connection, but it is only partial progress. To me this seems like looking at someone with a runny nose and fever – blaming the symptoms/evidence on the condition "having a cold" and not thinking one step further to contemplate dealing with the virus that is the ultimate cause. Likewise, many researchers commenting on magnetic field changes fail to look a step further and realize the link to pole shifts/crustal displacements that may be the ultimate cause of such quantum shifts in evolution.

When the populations of many species are decimated by a catastrophe that alters the arrangement of land and sea, this also cuts off and separates small communities and leaves some on new and virtually uninhabited lands – where any pre-existing abnormal traits in the genetics of the group that are beneficial traits are likely to spread out. New abilities from these traits, coupled with less competition, different climate, access to different foods, and many other changes, could often lead to two or more species branching off from a common ancestry. And when the next pole shift creates a new land bridge to connect formerly separate areas – the inhabitants mingle, and the eternal battle of the survival of the fittest continues.

Pole shifts causing sudden changes in climate, and the sudden separation and reintroduction of species, offer the best explanation for the extreme variation in the rate of evolution as well. Long periods of little to no evolution are separated by periodic bursts of "explosive" or "quantum evolution" which one paleontologist called a "virenzperiod." One species can suddenly give rise to dozens of subspecies. Such "adaptive radiation is episodic."[130] And as Darwin noted, perfectly well adapted species that survived for many ages can suddenly go extinct for no apparent reason, only to be replaced by inferior species that do not fill their niche as well. New species often seem to appear out of nowhere. But if pole shifts force life to migrate from one continent to another, if they constantly bring lands up and down, north and south, under the sea and back above it, we cannot expect to find evidence of all the "missing links."

Louis Agassiz

Louis Agassiz, a Swiss protégé of Cuvier's who later became a Harvard professor, wrote that: "All mountains and mountain chains have been upheaved by great convulsions of the globe, which rent asunder the surface of the earth, destroyed the animals and plants living on it at the time, and were succeeded by long intervals of repose… a time of building up and renewal followed the time of destruction." He noted the abrupt change in climate that had preserved mammoths and other carcasses so well: "A sudden intense winter, that was to last for ages, fell upon our globe; it spread over the very countries where these tropical animals had their homes, and so suddenly did it come upon them that they were embalmed beneath masses of snow and ice, without time even for the decay which follows death."[131]

Agassiz found evidence for glaciers spreading out from central points all over the world, including equatorial regions like the Amazon River Valley, Africa, Asia, Australia, and of course – North America. Yet he realized that the entire world had not frozen over at the same time. He wrestled with the idea that there was no clear explanation for "ice ages" (aside from the pole shift theory, which suggests that Greenland and Antarctica are in an "ice age" now because they are temporarily located at the poles) and noted in 1837 that if the earth remains in a fixed position as most people assumed, then we have no known trigger for ice ages: "We have as yet no clue to the source of this great and sudden change of climate. Various suggestions have been made – among others – that formerly the inclination of the earth's axis was greater..."

Frederik Klee

Frederik Alexander Gottlieb Klee published Le Deluge: Considération Géologique et Historique sur les Derniers Cataclysmes du Globe in 1847. Klee cited a theory by an abbot named Pluche, who "claimed that the deluge was caused by a displacement of the earth's axis accompanied by a subsidence of the earth's surface" in a book called Spectacle de la Nature, published in 1739. In chapter 12 of Le Deluge, Klee suggests that a "deplacement au l'axe du globe" (displacement of the earth's axis) explains many mysteries, including the well-preserved mammoths found in northern Siberia:

"...following a displacement of the earth's axis... The water, which suddenly turned to ice, completely preserved them from contact with outside air, and consequently, also from decomposition, which would have occurred after several days if they had remained in water or if they had been subject to the air."

Klee noted several previous equators, including when "the equator traversed the current poles, passing at around 90 degrees from the meridian of the

Faroe Islands. It was situated between Africa and Oceania, divided Asia and North America, and passed just to the west of South America." Klee realized that the often changing "direction of the earth's axis... explains satisfactorily the frequency and tremendous development of tropical plants and animals in current climates where we would least expect to find them..."

"By attributing to the earth's axis the direction as mentioned above, it will not be necessary to explain the exuberant development of the organic world by resorting to an arbitrary assumption that a different climate existed. My proposed hypothesis explains how tropical animals and vegetation, which always require warm climates, were formerly able to exist in areas which are currently subject to freezing temperatures for months on end. This includes how the elephant, the rhinoceros, the hippopotamus, the hyena, the tiger, and other animals of the pre-diluvian world were able to exist in England, France and Germany , etc.

Furthermore, why is it that the mammoth or the pre-diluvian elephant who, based on their structure must have lived in areas covered in rich vegetation, are mainly found in Siberia... following a displacement of the earth's axis, these lands were situated much closer to the North Pole, we would not be surprised to discover rhinoceros or mammoth cadavers in the glaciers of Siberia, where they must have been buried for thousands of years after having been seized by the flood in the regions they inhabited. This was a necessary consequence of the tilting of the earth's axis."

Bravo to Mr. Klee, who realized very early on that a pole shift had made Siberia colder at the same time that Europe and North America had grown warmer.

John Lubbock

In 1848 John Lubbock presented a paper to the Geological Society of London called "On Change of Climate Resulting from a Change in the Earth's Axis of Rotation." His major concerns involved centrifugal forces and the seawater comprising most of the Earth's equatorial bulge. He realized that a pole shift would bring some lands up from below sea level as they rotate away from the former equator, and that some lands would submerge beneath the waves like the myth of Atlantis as they move towards the new equator and are covered by the greater diameter of ocean water at the new equatorial bulge. It is not that the solid land itself does not change altitude at all; but the bulk of the changes in surface levels is so much more easily "fixed" (equilibrium is reestablished) by the flowing liquid of the oceans. Because the water readjusts its position so much faster, land altitudes relative to sea level experience tremendous changes as a result of a pole shift. The safest lands most likely to experience minimal loss of life must be those few that experience minimal changes in latitude or elevation – the pivot points of the pole shift. Other areas that begin and end at similar latitudes and elevations will also occur at middle latitudes but are harder to predict.

(Fortunately for our own odds of survival, it has been determined that most of the equatorial bulge is not seawater or any other part of the surface crust of the earth – the bulge is mostly in sub-crustal magma under the lithosphere.)

John Evans

Sir John Evans was the President of Britain's Geological Society, and Treasurer of the Royal Society, which was the most prestigious scientific society in Great Britain. In 1866 he published a paper titled "On a Possible Geological Cause of Changes in the Position of the Axis of the Earth's Crust." It goes without saying that he saw evidence for at least slow polar wandering, and possibly for rapid pole shifts, including "dislocations and undulations in the various strata are results which might be expected from the crust of the earth having to assume a new external form, if caused to revolve on a new axis." Evans should be credited as one of the early fathers of the pole shift theory, yet he is largely unknown.

Evans was trying to explain why "disturbances may lead, if not to a change in the position of the general axis of the globe, yet at all events to a change in the relative positions of the solid crust and the fluid nucleus, and in consequence to a change in the axis of rotation, so far as the former [the crust] is concerned." On the topic of tropical plant and animal remains frozen in Arctic locations, he suggests that with a pole shift "I think that possibly we may have here a vera causa such as would account for extreme variations from a Tropical to an Arctic temperature at the same spot, in a simpler and more satisfactory manner than any other hypothesis." "A change in the axis of rotation... no other hypothesis can well account for the existence of traces of an almost tropical vegetation within the Arctic circle.... [frozen remains] seem to afford conclusive evidence of a change in the position of the pole since the period at which they grew, as such vegetation must be considered impossible in so high a latitude."[132]

Alfred Drayson

Alfred Drayson was an Major General in the English Army, a friend of Arthur Conan Doyle, (Drayson is believed to have provided inspiration for Sherlock Holmes' adversary, Dr. Moriarty) an elected Fellow of the Royal Astronomical Society, and an author of dozens of books, including interesting titles like: <u>Great Britain Has Been and Will Be Again Within the Tropics</u> (1859) and <u>On the Cause, Date, and Duration of the Last Glacial Epoch of Geology, and the Probable Antiquity of Man: With an Investigation and Description of a New Movement of the Earth</u>. (1873) No one in recent decades seems to have read these books or have any copies, but the books are legitimately listed in many bibliographies and seem to indicate that Drayson was an early pole shift supporter. As a person, he was highly respected – though his theories on the "second rotation" of the Earth were not accepted by other scientists.

Charles-Etienne Brasseur de Bourbourg

The French historian and archeologist Charles-Etienne Brasseur de Bourbourg was another early proponent of such theories. He travelled Central America as a Catholic missionary and studied what little remained of Mayan culture after the Spanish Inquisition had burned most of their books. In 1861 he published a French translation of the Mayan creation epic, the <u>Popul Vuh</u>. In 1872 he wrote the "<u>Chronologie historique des Mexicains</u>" and described several global cataclysms caused by sudden shifts of the Earth's axis which he determined had started approximately 10,500 B.C. He wrote that "the disasters were caused each time by a shift in the axis of the world, upsetting the polar ice caps and reversing the order of the seasons."

Ignatius Donnelly

Ignatius Donnelly was a lawyer and a Congressman with varied intellectual interests. Some have suggested that he ran for Congress just to live in Washington D.C. and have access to the Library of Congress. In 1888 he published a book called The Great Cryptogram, in which he put forth the idea that Francis Bacon was Shakespeare. But in regard to pole shifts we are more interested in his best-selling book: Atlantis: The Antediluvian World, which came out in 1882, and to a lesser degree the follow-up companion book Ragnarok that came out a year later. Plato's tales of Atlantis were suddenly taken much more seriously again after Donnelly's attempt to factually establish that a great cataclysm really had destroyed this advanced ancient civilization and that Egypt, Mexico, and many other lands had cultures seeded by the survivors.

William Warren

Dr. William Fairfield Warren studied myths from around the world and noticed "how often the story of a falling sky and great flood was to be found intertwined with accounts of a lost island paradise. He also realized that the lost land had many polar features."[133] Warren wrote: what if "in consequence of the on-coming of the Ice Age, the survivors of the Flood were translocated from their antediluvian home at the Pole to the great central Asian plateau… the new aspect presented by the heavens in this new latitude would have been precisely as if in the grand world-convulsion the sky itself had become displaced, its polar dome tilted over about one third of the distance from the zenith to the horizon. The astronomical knowledge of those survivors very likely enabled them to understand the true reason of the changed appearance… in connection with some appalling natural catastrophe or world-disaster." Unfortunately Warren focused on the North Pole, not Antarctica (where there is actually land under the ice) because he believed one detail in the myths stated that the original homeland had submerged beneath a polar ocean, whereas Antarctica still had land above sea level, even if most of it lies beneath an ice sheet. In 1885 Warren wrote: Paradise Found: The Cradle of the Human Race at the North Pole.

Madame Blavatsky

The Secret Doctrine is a book published in 1888 by Helena Petrovna (Madame) Blavatsky, who had founded an occult organization known as the Theosophical Society in 1875. Whether a genuine psychic or just a wildly successful con artist, this woman allegedly was born into Russian nobility, travelled the world and developed her spiritual powers in Tibet. She eventually moved to New York, and accumulated nearly 100,000 followers after writing a few occult books on "Theosophy" or "Divine Wisdom." If nothing else, Blavatsky promoted many radical ideas including pole shifts.

She wrote about a "geological disturbance of the Earth's axis" repeatedly throughout the millennia; and that entire continents could be flooded and destroyed. She said that "Atlantis was the fourth" civilization destroyed by a change in the Earth's axis.[134] Blavatsky elaborated that there was a "great Deluge that carried away the Atlanteans and changed the face of the whole Earth (because 'the Earth [on its axis] became inclined.')"[135] She also wrote that "glacial periods" are due to "the shifting of the Earth's axis – a proof of which may be found in the Book of Enoch... when speaking of 'the great inclination of the Earth...[which] is in travail.'"[136] According to her, in recent human history "there have already been four such axial disturbances"[137] including one that destroyed the civilization of Mu or Lemuria. She said that modern scientists "may pooh-pooh the idea of a periodical change in the behavior of the Globe's axis, and smile at the conversation given in the Book of Enoch between Noah and his 'grandfather' Enoch, the allegory is nevertheless a geological and astronomical fact. There is a secular change in the inclination of the Earth's axis, *and its appointed time is recorded in one of the great Secret Cycles*."[138]

Marshall Wheeler

Marshall Wheeler wrote a book in 1889 called The Earth – Its Third Motion, in which he said "The Earth will come to rest at 90 degrees from its starting point, reversing the present position of the poles and [two points on] the equator." He closed with: "...the whole world should be informed of the fact, and means taken to forever perpetuate that knowledge, so that when the dread event transpires, mankind should not lapse again into prehistoric barbarism." We may rest assured that Wheeler is not the only one with these thoughts.

The Roman historian Flavius Josephus wrote the Antiquities of the Jews around the year 94 A.D. He described two pillars carved by Noah's grandfather Enoch, one of marble and one of brass, to resist the effects of fire and water... On them were carved the history of creation, the knowledge of arts and sciences, and the speculative mysteries of religious beliefs. "That their inventions might not be lost before they were sufficiently known, upon Adam's prediction that the world was to be destroyed at one time through the force of fire and at another time the violence and quantity of water, they made two pillars..." Josephus also wrote that near the pillars was buried a sacred treasure of knowledge in a great subterranean vault. (Many New Age followers of Edgar Cayce, mentioned in detail later, suspect that Cayce's prediction that an Atlantean Hall of Records can be accessed from an entrance buried near the front paws of the Sphinx, were given a boost when such a hollow chamber was discovered by sonar readings in the 1990s. Unfortunately, Egyptian authorities have forbidden any digging, entrance, or further investigation of it.)

Around the year 850 A.D. the Persian astronomer Abu Ma'shar al-Balkhi wrote that Egypt's pyramids were "built to protect scientific wisdom from before the Flood... including engraved explanations of science, in order to pass them on to those who would come after." Another Middle Eastern writer, Al-Masudi, wrote The Meadows around 947 A.D. and commented at length on the pyramids as the pillars of Enoch, which were in ancient traditions built of different materials in pre-diluvian times to record ancient wisdom in a way that would withstand both fire and flood. Al-Masudi wrote: "They had learned from a study of the stars that a catastrophe threatened the land, but they were uncertain whether the world was to perish by fire [or] by a deluge... In fear lest the sciences should be annihilated with the people, they constructed these berabi and disgorged their knowledge into the figures, the images, and the inscriptions."

In modern times, the Stelle Group, founded by Richard Kieninger in 1963, set up a town (Stelle) in Illinois exclusively for members of their group; a group that believed themselves "to be among an intended 'remnant' and is following an ancient mandate... to [re]establish the Nation of God on Lemuria after the coming pole shift... this 'geological reapportionment' is a predictable, natural occurrence foreseen long ago by the Brotherhoods, and they have included it in their plans..."[139] As I said, Marshall Wheeler is far from the only one concerned about maintaining humanity's knowledge and progress after the pole shift.

James Churchward

In 1926, British engineer and occult writer James Churchward published <u>The Lost Continent of Mu Motherland of Man</u>. It was the first of his many books on Mu, which he claimed was a lost continent in the Pacific Ocean that stretched from near Hawaii towards Fiji and Easter Island. He claimed it was the original Garden of Eden, with millions of inhabitants and technology more advanced than that of his day in the early 20th century. He believed that Mu had colonized Atlantis and the Americas, then Egypt and other parts of the Middle East, Africa, and Europe "until cataclysms destroyed Mu and Atlantis."

Churchward claimed that while a British military officer in India, he had developed a long friendship with a Hindu high priest who taught him ancient Vedic knowledge and showed him very ancient artifacts including clay tablets and texts that spoke of an ancient global civilization that disappeared after cataclysmic earth changes. Churchward was so convinced of the truth of these Hindu stories of lost civilizations that he devoted many years to finding additional evidence, travelling to Egypt, Tibet, Mexico, and especially the South Pacific, where he placed the lost continent of Mu. This is somewhat strange to do, since Augustus Le Plongeon had first written about Mu as an Atlantic island continent, and had based his thinking off a Mayan reference in the Troano Codex to "a land which had been submerged by a catastrophe."

Churchward wrote that Mu "was completely obliterated in almost a single night" by earthquakes, volcanic eruptions, and a flood that submerged it under "millions of square miles of water." Douglas Kenyon, David Hatcher Childress, and other authors writing introductions to modern editions of his books describe Churchward's destruction of Mu as occurring during a "pole shift" but I can't find the phrase put exactly that way in any of Churchward's own words. Even so, he definitely helped promote the concept of ancient civilizations lost to global natural disasters. (Childress has written extensively on Mu/Lemuria and suggests that it was an advanced world power from 78,000 to 26,000 years ago, when he suggests it "sank in a cataclysmic pole shift."[140] Assuming I am correct that pole shifts occur in cycles of just under 13,000 years, this would place the destruction of Mu or Lemuria two pole shifts back, and its foundation would have been six pole shifts back. Churchward assumed Mu was destroyed with the last pole shift, 12-13,000 years ago.)

Edgar Cayce

Edgar Cayce, America's "Sleeping Prophet," may or may not have been an accurate prophet. He made too many inaccurate predictions about earthquakes, the rediscovery of evidence of Atlantis, and a pole shift which did not occur on schedule around 1998-2001 when he said there would be "upheavals in the Arctic and Antarctic... the shifting then of the poles."[141] On more than one occasion he said this same time frame would be "when there is a shifting of the poles. Or a new cycle begins."[142] But I must mention him because he was a famous and early proponent of the pole shift hypothesis, with a bit of a twist: while Cayce acknowledged the devastating changes in latitude that pole shifts cause, he focused more than most on changes in elevation and altitude, with many comments about: "areas to rise; while many a higher land will sink."[143] He also said: "As to the changes physical again: The Earth will be broken up in the western portion of America. The greater portion of Japan must go into the sea. The upper portion of Europe will be changed as in the twinkling of an eye. Land will appear off the east coast of America."[144]

On past pole shifts, he commented that one occurred almost 13,000 years ago; that a surviving colony from Atlantis rebuilt civilization in Egypt, and he even attributed the cause of the pole shift to ice. Years before this cause was popularly theorized by Brown or Hapgood, Cayce said while in one of his trances that "ice... changed the poles."[145] He may even have attributed a related cause from cycle in the Sun and waves of energy that periodically reach us from the galactic center: "the passage of the Sun through other spheres of activity that creates this solar cycle." This would be in line with thoughts from Dr. Robert Schoch and Dr. Paul LaViolette, but despite finding this last Cayce quote online, I can't find a reading number or date when Cayce actually said this...

Cayce allegedly discovered an ability to enter a trance state much like sleep, during which he would speak with great conviction, and apparent otherworldly knowledge – on topics from medical cures to accessing past life experiences in ancient Egypt and Atlantis to viewing future events. He repeatedly spoke of Atlantis (destroyed in a previous pole shift) and an upcoming pole shift due near the turn of the millennium. He gave detailed descriptions of new coastlines and lands that would be safe after a shifting of the earth's axis, even including a detailed description of his future life reincarnated in 22nd century Nebraska – which would then be on the new coast of the Pacific Ocean.

Despite predictions which have failed to occur long after his death in 1945, Cayce was revered during his life and especially at the peak of interest in New Age topics in the 1960s and 1970s. Books about his pole shift predictions may have brought the idea into the American psyche more than anyone else had done. And his prophecies are not completely without merit.

One interesting comment of Cayce's is that survivors of Atlantis, over 12,000 years ago, settled in just a few areas: Egypt and Israel, the Basque region in the Pyrenees between Spain and France, the Gobi Desert in Mongolia (which he says was fertile then) the Yucatan Peninsula of Mexico where the Maya still live, and the Iroquois region of eastern Canada. This is a strange set of apparently unrelated areas for alleged Atlantean migration.

But Cayce's comments on this matter saw an interesting revival in 1997 when advances in genetics revealed that although the mitochondrial DNA of humanity falls into four main haplogroups, they had just discovered a fifth less common Haplogroup X. This particular mtDNA is very rare, but noticeably present in a few scattered regions: the Basque region of Spain, Egypt and Israel, the Gobi desert, and the Iroquois region of eastern Canada – where it was substantially introduced approximately 10,000 B.C. These findings may very well back up the idea that Atlantis' survivors introduced their distinct genetics into the very regions Cayce mentioned...

I recently stumbled onto an article from 2009 by Donald Yates titled: "Anomolous Mitochondrial DNA Lineages in the Cherokee" which notes certain genetic markers prevalent in both the Cherokee Indians and in Egypt and Israel, and comments "The geneticists always seem to cry 'post-Columbian admixture' but fail to take into account that there are no plausible post-Columbian sources for the particular genetic mix encountered."

Another vision of Edgar Cayce's caught my attention for its apparent foolishness when I first read it around the year 1980. I have tried in vain to find it again more recently, but despite being certain I have read it, I cannot tell you what book I read it in; my recent searches have been fruitless... But I clearly remember reading about Cayce describing a vision in which he was standing in a city with rubble and dust everywhere, as if the city had been at least partially destroyed by earthquake or war. When I read it as a boy in 1980, influenced by duck and cover drills at school in a time when most Americans expected an eventual nuclear war with the Soviet Union, I interpreted the vision to describe the aftermath of a nuclear attack.

Cayce said that in this vision he walked up to a man in a bulldozer wearing a dust mask and asked him where they were. The bulldozer driver replied: "C'mon buddy! This is Manhattan!" At the time, I thought this was ridiculous, knowing full well that a dust mask would be the least of their concerns if New York had just been nuked. But roughly September 13, 2001... I got the eerie feeling that Edgar Cayce could have been experiencing some sort of astral projection to New York to witness the rubble of the Twin Towers collapse...

Whether an accurate psychic or a fraud or somewhere in between, Cayce definitely brought the topics of Atlantis, a previous pole shift, and an upcoming pole shift into the minds of millions of Americans like no one had before. His prophecies of pole shifts have inspired many other authors and books, especially William Hutton and Jonathan Eagle's <u>Earth's Catastrophic Past and Future: A Scientific Analysis of Information Channeled by Edgar Cayce</u>.

Hugh Auchincloss Brown

The first American scientist to achieve fame for supporting the pole shift theory was an electrical engineer named Hugh Auchincloss Brown. He started work on his theory in 1911 after reading about the discovery of frozen mammoths in the Siberian tundra, and continued to periodically revise his thoughts over the next few decades. Brown first published his theory on recurring pole shifts in Time Magazine in 1948, with an article titled: "Can the Earth Capsize?"

Brown suggested that the imbalanced accumulation of ice at the poles (about 293 cubic miles, or twenty billion tons of ice per year) causes increasing imbalances in centrifugal forces on the planet (not just the outer crust) and that this causes a tipping of the axis of rotation of about 80 degrees, bringing the former poles to rest near a new equator approximately every 6,500-7,000 years. Brown believed that we are due for the next catastrophe "soon" – possibly by the end of the 20th century or the 21st century.

He was very concerned over "the certainty of a future world cataclysm during which most of the earth's population will be destroyed in the same manner as the mammoths of prehistoric times were destroyed." Brown explained: "Because of the curvature of the globe, the centrifugal forces of the rotating ice sheets which initiate the careens soon reach a maximum and then diminish. When the ice caps have migrated 45 degrees of latitude, their centrifugal force responds to the combined motions of careening and rotation. Between the sun latitudes of 45 degrees and 0 degrees they change from being upsetting to being stabilizing forces. Equatorial forces then start to form, and the centrifugal forces of the ice caps and of the new bulges are soon working in unison to bring the reeling motions of the globe to a rapid slow-down and stop..."[146]

Brown assumes that East Antarctica will move north to the equator along the meridian of 135 degrees east longitude.[147] "Following the next careen of the globe the present continent of Antarctica can reasonably be expected to become the center of a land hemisphere --- because of the centrifugal force of rotation which shall be created by its weight and speed of motion... The area of the globe now occupied by the Arctic Ocean will probably become the center of a water hemisphere --- like the Pacific Ocean today. What is now northern Siberia, northern Canada and Alaska will probably become parts of the submerged ocean floor... "According to this hypothesis, Brazil would roll to the South Pole and the Philippine Islands would become the land area nearest the North Pole... It would be equally valid speculation to say that some area of the globe within about 2,000 miles of Lake Chad will be at the North Pole during the epoch of time following our own, and that this would occur as a result of the past --- namely, the Hudson Bay Basin careened to the North Pole axis of spin, then Lake Chad moved in, only to be supplanted by the present Arctic Ocean area. This shows a tendency for land areas to roll back to nearly the same position of latitude and longitude that they rolled away from."

Brown wrote: "The former North Pole area included Egypt... Egypt escaped the Flood, for at the 'dawn' of Mediterranean history the nation appears the be mature, old... the civilization of ancient Egypt, upon its first appearance, was of a higher order than at any subsequent period of its history... it was mature when it first appeared."[148] Although I agree that Egyptian civilization started off advanced, and seems to have been started by the survivors of a previous civilization - I am not convinced that there was a pole shift approximately 6,500-7,000 years ago or that the Sudan Basin around Lake Chad was the site of the North Pole at any time in the last 50,000 years.

Brown concluded that pole shifts happen at irregular intervals including 7,000 years ago, 4,400 years before that, and 5,000 years before that – which does not align with the regularity of any portion of the cycle of precession which other researchers conclude is linked to pole shifts. "The Egyptians viewed their living universe as a rhythmic movement"[149] of regular and repeating astronomical cycles and patterns. Brown's Irregular timing does not match the Egyptian description of "the usual interval" or the regularity of India's cycles of Yuga Ages or China's Ten Ages... A polar ice cap covering Egypt from 11,400 years ago to about 6,000 years ago also does not seem to fit any geological, zoological, or archeological evidence. But I feel obliged to mention the details of Brown's conclusions anyway, as he is one of the central figures in the modern development of the pole shift theory. This also helps to demonstrate that despite the same world of evidence in favor of periodic pole shifts – different scientists reach different conclusions.

Brown suggests that we may be experiencing one of the longest epochs of stability in history, and that this time is different because there is no land mass at the North Pole and only one permanent ice cap, allowing us to enjoy a longer than average civilization already. We normally have two ice caps, north and south, but warm Pacific waters flow through the Bering Strait and mitigate the cold temperatures in the Arctic Ocean as they cycle through towards Greenland. Brown suggests: "Were it not for the continuous flow of water, from west to east through the Bering Strait, the Arctic Ocean – which averages 4,000 feet in depth – would have frozen solid centuries ago, and an Arctic Ice Cap would have developed. With two Ice Caps – one in Antarctica and another in the Arctic – our epoch of time, between two careens of the globe, would have been radically reduced in length, and our present civilization would not have had a chance to develop."[150]

Brown kept working on these ideas, and in 1967 published a book called <u>Cataclysms of the Earth</u>, in which he warns us that: "The present ice cap in Antarctica is merely the last of many thousands that have previously existed. Geological records reveal that it is the successor to a long lineage of glistening assassins of former civilizations."[151] On the topic of mammoths in Siberia he wrote: "just prior to the last careening of the globe this region was populated with teeming herds of animals. They lived there because an ample food supply existed, and the food supply grew because the climate was warm."

Brown suggested that pole shifts occur with great speed in a single day, that the entire body of the planet rotates, and that the poles move about 75-80 degrees, but probably not a full 90 degrees latitude – to stop near the equator. "Because of the curvature of the globe, the centrifugal forces of the rotating ice caps which initiate the careens soon reach a maximum and then diminish… the new bulges of the earth are soon working in unison to bring the reeling motions of the globe to a rapid slow-down and stop. In the meantime, kinetic energy which has developed in the continental land masses because of their weights and velocities, collides with the combined energy of the moving continents is absorbed by the crushing, elevating, and wrinkling of large land areas whose rock strata are crumpled and bent in ridges"[152] and mountains… Extra mass near a pole is destabilizing; extra mass near the equator is stabilizing. When the imbalance of mass that starts a pole shift gets close enough to the equator, it helps to slow down and end what it started.

Brown says that before the last pole shift, the North Pole was in the Sudan Basin near Lake Chad until about 7,000 years ago, and that the previous North Pole before that had been in Hudson Bay about 4,400 years earlier. I see little other evidence for a North Pole situated in the Sudan at this time, and other prominent pole shift researchers like Hapgood do not share the belief that the area was arctic just 7,000 years ago. But there is evidence for previous North Poles in "NE Africa"[153] much further back in time (assuming Brown is wrong merely on timing, this could explain the evidence that mislead him) and I will admit that a recent North Pole near Egypt – even though I do not think there was a recent one – could explain the "Epipaleolithic–Predynastic Gap" in which there is little to no evidence for civilization in Egypt between approximately 12,500 to 5,100 years ago. On the other hand, "massive floods 12,500 years ago… [followed by the end of wet and fertile conditions throughout Egypt replaced with] …the present dry period"[154] could also be the explanation for the gap, and could have resulted from the pole shift positions described by Hapgood.

Whether Brown had all the details of his pole shift theory right or not, he made a point of warning reporters, scientists, and politicians about his conclusions. He felt the problem was urgent, because we had already exceeded what he believed was the average length of stability between cataclysms. Brown suggested spending millions of dollars to investigate the evidence and determine the best ways to forestall the pole shift, which he suggested could be done using atomic bombs to reduce the Antarctic ice cap. "We are faced with the alternative of limiting the growth of the present ice cap or accepting a limit to the duration of our present epoch…. We can postpone the fatal day when the great Ice Cap will roll the globe sideways."[155]

I should emphasize that Brown believed that growing ice caps and the imbalanced centrifugal forces they impart to the crust of the earth are the predominant cause of pole shifts. Yet he did acknowledge that the straw that breaks the camel's back – the trigger that could push earthly imbalances into action – are from "cosmic forces of extraterrestrial nature."[156] He believed that cosmic electromagnetic fields power both gravity and rotational spin, and suggests that a pole shift "may start because of changes in the directions and intensities of the forces of celestial radiation" and "the probability of the existence of a super cosmic cycle in the eternity of time, involving sun, earth, moon, and possibly other celestial bodies... Cosmic Geophysics."[157]

Brown was largely scorned by his contemporaries. But he may have indirectly gotten his concerns through to the upper levels of government. We have a seed vault in Norway's Svalbard Islands, almost 80 degrees north latitude, housing hundreds of thousands of seed samples to maintain crop diversity in the event of a global crisis. Hmmmm. We have thousands of D.U.M.B.s – Deep Underground Military Bases – the one in Missouri has hundreds of miles of tunnels filled with supplies the government may need after some unknown catastrophe they are preparing for. They could just be concerned with nuclear war, or a pole shift, or who knows what. But they are making a huge effort to prepare for something.

Immanuel Velikovsky

Immanuel Velikovsky wrote a book called <u>Worlds in Collision</u>, published by MacMillan in 1950. It talked about ancient myths of catastrophe and a possible cosmic trigger for pole shifts, which Velikovsky described in violent detail: "Let us assume, as a working hypothesis, that ...the axis of the earth shifted or tilted. At that moment an earthquake would make the globe shudder. Air and water would continue to move through inertia; hurricanes would sweep the Earth, and the seas would rush over continents, carrying gravel and sand and marine animals, and casting them onto land. Heat would be developed, rocks would melt, volcanoes would erupt, lava would flow from fissures in the ruptured ground and cover vast areas.

Mountains would spring up from the plains and would climb and travel upon the shoulders of other mountains, causing faults and rifts. Lakes would be tilted and emptied, rivers would change their beds; large land areas and all their inhabitants would slip under the sea. Forests would burn, and the hurricanes and wild seas would wrest them from the ground on which they grew and pile them, branch and root, in heaps. Seas would turn into deserts, their waters rolling away.

...the water confined to the equatorial oceans by centrifugal force would retreat to the poles, and high tides and hurricanes would rush from pole to pole, carrying reindeers and seals to the tropics and desert lions to the Arctic, moving from the equator up to the mountain ridges of the Himalayas and down the African jungles; and crumbled rocks torn from splintering mountains would be scattered over large distances; and herds of animals would be washed from the plains of Siberia.

The shifting of the axis would change the climate in every place, leaving corals in Newfoundland and elephants in Alaska, fig trees in northern Greenland and luxuriant forests in Antarctica. In the event of a rapid shift of the axis, many species and genera of animals on land and in the sea would be destroyed, and civilizations, if any, would be reduced to ruins."

Worlds in Collision topped the New York Times best-seller list for 11 weeks, and stayed in the top ten for 27 weeks. It was sensational, popular, and utterly unscientific. I almost didn't want to mention him in this book, because his inventive mechanisms for the causes of such catastrophes are so ludicrous. The scientific community was so against Velikovsky's nonsense that they told MacMillan Publishing if they don't stop publishing the book, no one will buy their scientific textbooks and their role as a publisher for scientists, universities, and schools will be flushed down the toilet. MacMillan gave in and transferred their rights to the book to another publisher while it was still number one on the charts. Sometimes you can be right (pole shifts have happened and will happen again) for the wrong reasons (Jupiter did NOT give birth to Venus around 1400 B.C., and Venus did not fly past Earth and cause a pole shift before assuming its current orbit closer to the Sun.)

Quite possibly due to the popularity of Velikovsky's book, the 1950s saw greater interest in the idea of cataclysms and pole shifts. K.A. Pauley wrote in "The Cause of the Great Ice Ages"[158] that "the lithosphere was displaced during the great Ice Ages, and that the displacements [of the surface of the Earth] were the direct cause of the alterations in climates." S.K. Runcorn wrote in Scientific American in 1955 that "the planet has rolled about, changing the location of its geographical poles."

Velikovsky may not have made sure that all his theories had the slightest chance of being scientifically valid before writing a book, but he certainly knew how to sell books and make money. In 1955, an updated version of Worlds in Collison came out under a new title: Earth in Upheaval. Both books are still selling and being quoted today. But as foolish as some of Velikovsky's claims may be, he assembled a good amount of mythology and scientific evidence that pole shifts had happened in the near-historical past, and he contributed greatly to the public awareness of the idea of pole shifts: "This catastrophic shifting of the axis, once or a number of times... have actually taken place... The evidence is... overwhelming that the great global catastrophes were either accompanied or caused by shifting of the terrestrial axis."[159]

John White, in his book Pole Shift asks: "Does the Bible prophesy a pole shift? Does it contain evidence of a previous one? ...[One which some suggest] could have only occurred due to a sudden tilting of the earth's axis. Moreover, they point to passages in both the Old and New Testament that indicate, from their perspective, than an axis shift will again take place..."[160] A few pages later White paraphrased Velikovsky's thoughts on Bible prophets: "The cosmic catastrophes they foresaw were predictable by them on the basis of their astronomical observations and their knowledge of previous cosmic catastrophes..." This is an idea I wholeheartedly endorse – and discuss at length in previous books of my own.

But I think Velikovsky's ideas about Venus and Jupiter are just silly. Venus obviously existed prior to 1400 B.C. and was recorded by many ancient cultures. There is no scientific explanation that would allow one planet to burp out another. But because I suspect that the most likely trigger for pole shifts is the series of periodic outbursts coming from the galactic center (as discovered by Carl Seyfert around 1940, and understood better by Dr. Paul LaViolette around 1980) I do agree with Velikovsky when he says: "We cannot imagine any cause or agent for this, unless it be an exogenous agent, an extraterrestrial cause. For the removal of the poles from their places, or the shifting of the axis, also, only an external agent could have been responsible."[161] I say this because it seems that mass imbalances of an off center ice cap, with or without additional mass imbalances beneath the crust, are not close to the point where they could shear the crust loose from the layers below – not with the geomagnetic field at current strength, anyway. (It is a very different equation if the Earth's magnetic field is substantially weakened.)

One high ranking politician was taking someone's warnings on this topic seriously; and admitted a concern about a pole shift in 1956. Estes Kefauver was no dummy, so I take his comment seriously. He was a math teacher before graduating from Yale Law School. He practiced as a lawyer for years, and was a Congressman from 1939-1949 and a Senator from 1949-1963. The Democratic Party chose him as their Vice-Presidential candidate in 1956, but the Stevenson–Kefauver team lost to the reigning Eisenhower–Nixon team in the 1956 election. Kefauver remained a powerful Senator after the national election loss to the extremely popular incumbent president General Dwight Eisenhower, who we must remember had won WWII against Germany in the eyes of most voters – there was no shame in losing that election. Kefauver was named chair of the U.S. Senate Antitrust and Monopoly Subcommittee in 1957 and served as its chairman until his death in 1963. I point all this out to emphasize, Kefauver was not a crackpot or an idiot. If he mentioned a pole shift, he had a reason to worry about one.

Various newspaper articles around October 17-19, 1956 tell us that Kefauver addressed the Atomic Energy Commission and called for a ban on underground Hydrogen Bomb testing, suggesting, among other potential problems it could cause, that such explosions could help cause a pole shift and that the earth could be "knocked 16 degrees off its axis." This sounded so outlandish I had to confirm the first sources I read his quote in, but I found it again in <u>Estes Kefauver: A Biography</u>.[162] This 16 degree shift sounds very specific; as if based on something convincing he was told or given to read... What did he know? Who had spoken to him on this topic? Brown? Velikovsky? Adam Barber? Or maybe Hapgood and Einstein, who analyzed Brown's theory? Brown contacted all major politicians with a pole shift warning... I'd love to know who convinced Kefauver of this danger, and what information the 16 degree shift is based on...

Charles Hapgood

He may not be the first author to discuss it, but the best known proponent of the pole shift theory is Professor Charles Hapgood, who published his first book on his Earth Crust Displacement (ECD) theory of pole shifts in 1958 with the title: <u>Earth's Shifting Crust</u>. Though "merely" a historian and not a scientist, Hapgood was a highly intelligent, Harvard-educated man capable of in many areas of expertise. Before becoming a history professor he worked for the OSS, the office of Strategic Services that was the precursor to the CIA, and he served as the liaison between the White House and the Secretary of War during WWII. He was a well-respected man with good contacts.

A student in Hapgood's history class brought up Brown's theory on pole shifts, and although he first assumed that if the class investigated a little they would prove the theory was nonsense – it turned out to be a life-changing project for Professor Hapgood. Brown's theory had merit. In the course of the research that followed, Hapgood, who couldn't help looking at the problem from a historical angle – looked deeper into historical anomalies like ancient maps showing lands not yet discovered, as if European mapmakers were modifying very ancient world maps based on "new" discoveries after Columbus reached the New World – and found that the regions Europeans had not yet reached and should not have been on the maps of the time at all were often the MOST accurate sections, because primitive European explorers, with no good way to determine longitude, had not messed up the details from the ancient maps yet with more modern flawed observations and calculations...

Captain Arlington Mallory brought to his attention the possibility that the Piri Re'is map of 1513 showed the ice free coast of Western Antarctica below South America. Antarctica was not officially (re)discovered until 1820. Hapgood's investigation of maps soon had him wondering if pole shifts had occurred while human civilization had existed to map Antarctica and the rest of the world in its previous position, before the ice covered all of Antarctica...

Hapgood argues that the slow and gradual processes believed to shape the crust of the planet over hundreds of millions of years simply do not explain the evidence for repeated catastrophic changes in very short intervals of time. Ice Ages are a concept they completely fail to explain. Why, for example, is there evidence of ice ages all over the planet, even in India and Africa, when there is no evidence the entire planet cooled down enough to allow those lands to freeze at their current latitudes? How could tropical species like chimpanzees or coconuts or corals have survived a global freezing? Why do the glaciers of various ice sheets seem to spread out from their centers, instead of from the nearest pole?

Hapgood's analysis of the North American Ice Sheet shows it centered on Hudson Bay, and that when it formed the ice spread out from the center of the bay, not from our current North Pole then southwards. And if this ice sheet had formed when Hudson Bay was around 60 degrees north latitude like it is today, then why wasn't there an ice sheet at the same latitudes in Asia? The evidence in Russia during the last "Ice Age" suggests lush vegetation, dense forests, and huge herds of large animals. The most logical explanation is that Hudson Bay and the North Pole were at the same location, and the ice spread out from the new North Pole there after a pole shift had relocated Hudson Bay to the Arctic. After the next pole shift, which moved the crust again and brought the North Pole to the Arctic location we know today, Hudson Bay moved about thirty degrees south, away from the pole, and the North American ice sheet melted. However, Hapgood did not see evidence for a North Pole in the Sudan Basin as Brown claimed.

He also disagreed with Brown on the idea that the entire planet would rotate as a solid body during a pole shift. Hapgood teamed up with an engineer named James Hunter Campbell, who came up with the idea that if the mass imbalance of the ice caps did not produce sufficient centrifugal force to alter the spin of the mass of the entire planet, they would certainly have enough force to move just the crust of the earth, if the outer layer could detach and slide over the mantle beneath it, and he developed this theory in <u>Earth's Shifting Crust</u>.

Hapgood suggests that between pole shifts there are long periods when stresses gradually build up in the earth's crust. Unlike slowly tilting a container of liquid and having the contents readjust quickly and gradually, the earth's crust has significant friction keeping it from sliding across the layers below. Imagine tilting a board with a brick on it. The brick stays in place despite an increasing angle of inclination, until the imbalance suddenly overcomes friction and the brick completely loses its position and falls to the bottom. Hapgood also compares a crustal displacement (pole shift) to bending a tree branch, until it reaches the breaking point when it suddenly snaps and completely gives way completely to the forces exerted on it. In the case of a pole shift, forces actually increase after this point.

Once movement between layers begins, heat from the friction of movement further liquefies the semi-molten layers and reduces friction. And the tangential centripetal forces from the off-center mass of the ice sheet multiply as it starts to move further from the axis of rotation of the core. Like a wobbling top losing gyroscopic spin stability, the wobble increases as the top starts to fall. Ice sheets centered five degrees from the South Pole exert half the tangential centripetal forces that they have once they have reached a position ten degrees away, and double again at twenty degrees from the pole, increasing and accelerating as the imbalance of mass approaches the equator.

An important question, we quickly realize, regards this speed of movement. If the imbalanced ice mass near the pole is the primary cause, and the pole shift moves slowly over centuries or millennia, then the ice will melt, the imbalance of mass will be eliminated, and the magnitude of the pole shift will be just enough to move the ice caps into latitudes that will melt them – perhaps thirty degrees of latitude. But if the ice (or any other sub-crustal) mass imbalance is the main cause and the movement happens in just hours, days, or weeks – then the polar ice sheet does not melt away fast enough to stop the pole shift before the imbalance reaches the equator (possibly even overshooting it before reverting back.) Brown concluded that pole shifts occur in a single day, and end with the old poles and ice caps by the equator, with moves of about 80 degrees latitude. Hapgood wavers back and forth in his statements, making me wonder if he believed in a very fast pole shift, but felt he had to be conservative in his comments on speed in order to keep his job or have his overall ideas accepted. Once in a while he acknowledged sudden catastrophes like tsunamis and the quick death and freezing of entire herds of many species animals including mammoths... He occasionally made statements like: "shifts of the lithosphere have at times attained extraordinary speeds."[163]

But most of the time Hapgood stated that pole shifts are very slow events, taking thousands of years to complete, and that the total change in latitude is approximately thirty degrees in a typical pole shift. He did not focus on evidence of sudden catastrophes, but commented often on evidence of gradual forced migrations due to the slow but inevitable freezing or flooding of various ancient homelands – though he eventually condensed his time frame for a pole shift down to perhaps a few centuries. In Path of the Pole (p. 40) he cites an article by Ernst Deutsch ("Polar Wandering and Continental Drift") in which Deutsch argues for very slow pole shifts, then laments evidence Deutsch found in Scotland of a crustal displacement occurring at a speed "several orders of magnitude" faster than he thought it should happen. (Presumably thousands or tens of thousands of times faster.) Hapgood wrote: "The evidence points to a very rapid transit of the pole from its old to its new home. It must have completed its transition in a matter of centuries rather than millennia." But was he, even in that statement, being overly cautious to admit how fast some evidence indicates pole shifts happen? Why was he minimizing, in his own words, the "astonishing suddenness" of the evidence and dragging out the time frame of the pole shift?

Was Hapgood caving in to politics and peer pressure among academics who only accepted very slow changes that fit their uniformitarian models, while dismissing evidence of sudden catastrophes? In the updated version of Earth's Shifting Crust titled Path of the Pole Hapgood lamented: "Such is the frailty of the human mind, scientific or not, that displacements of the lithosphere have been pushed into the background, and all the attention has been paid to continental drift." (The biggest difference in the updated version of the book is that Hapgood originally argued that off-center ice caps would have a large enough mass imbalance to eventually trigger a pole shift – but in the updated version he says that new information on "the earth's crust now suggests that the forces responsible for shifts of the crust lie at some depth within the earth rather than on its surface."[164] Hapgood continues to discuss the imbalance of the ice cap's off-center mass distribution, but no longer claims this is the main or only cause: "Since the abandonment of the ice cap as the cause of the displacement of the lithosphere, the ice cap here can be taken to represent a net imbalance of the distribution of mass in the lithosphere."[165] But as to why Hapgood also went back and forth between "astonishing suddenness" and "centuries" and "very rapid" and "millennia" I cannot easily explain. A few pages later he wrote: "There was nothing in this [the melting of the North American ice cap] to suggest the painfully slow pace of usual geological history. To be blunt about it, it was a catastrophe, a cataclysm…" On the same page (p. 150) he says it took "little more than a millennium, perhaps two." On the next page he says "the movement might have continued for four thousand years." And on the next page, he explains why evidence indicates it was "necessary for the polar shift to have been completed very quickly." On page 162 he mentions the "rapid transit of the pole" as "fast enough so that the ice cap did not start growing at the coast and move inland; it started in the Hudson Bay region, and after it had grown thick enough to move, it spread outward in all directions. A very remarkable evidence of the suddenness with which the ice cap was born is the fact that it contained thousands (and perhaps millions) of animals of a temperate climate…" Why, Dr. Hapgood? Why did you so torturously obscure your true opinion on how fast pole shifts occur? I have to assume Hapgood was threatened and forced to tone down his claims by a government that does not want people to panic about an imminent pole shift. In 1982 when he was about to publish more sensational claims, he was hit by a car and killed. Accidents happen to people about to publish undesirable findings sometimes.

There is undoubtedly a combination of factors at play, including mass imbalances above and below the surface, along with a reduction in viscosity and friction between layers during a weakening or reversing magnetic field, and also a cycle of extreme changes in much larger cosmic fields.) Hapgood may have been torn between fast and slow pole shifts because he saw evidence for both. He saw evidence for ice caps melting away without reaching the equator, and shifts of about thirty degrees latitude – very strong evidence for a very slow pole shift. He also saw evidence of tsunamis and mass extinctions – very strong evidence of a very rapid pole shift. How can the two be reconciled? Hapgood eventually realized that the imbalanced mass of the Antarctic ice cap alone might never be sufficient to overcome the stabilizing friction between the crust and the liquid magma layer below, so he hypothesized an additional imbalance of mass below the surface much also be present. I think the answer may be that mass imbalance, above and/or below the surface, is not the only cause to consider.

I would like to make an analogy between the ice-cap's center of mass being hundreds of miles off-center from the rotational pole and a monster locked in a cage. If the monster gets out and causes destruction, we would all blame it for the deaths of its victims. But what let it out of the cage? Likewise, the ice mass imbalance is a very real problem and one day that mass will move toward the equator to correct the imbalance. But there has already been an imbalance for thousands of years; what will unlock the cage that allows it to move? Most researchers eventually conclude there must be a cosmic trigger that exerts some great force on the earth.

The answer may lie in the relatively new ideas of magnetohydrodynamics.

Hannes Alfven won the Nobel Prize in Physics in 1970 for developing this new field, which among other things, explains why some liquids like mercury and iron have very different properties when in a magnetic field. Specific to our concerns, liquid mercury in a container in no electric field disperses spin, and rotation in one part of the liquid does not cause spin currents throughout the liquid mercury. Place the same container in a magnetic field, and it suddenly does the opposite, transferring spin throughout. This is important to our pole shift theory, because liquid iron in the Earth's mantle does the same thing, acting viscous while in a magnetic field and transferring momentum through high friction between layers.

Liquid rock, the magma above the core, has substantial amounts of iron and other metals in it and is constantly flowing and moving like the insides of a lava lamp. As one article addressing the accelerating magnetic pole shift in early 2018 noted: "Under the Earth's lithosphere, or crust… there is a vast ocean of super-heated iron and other metallic elements surging around the core. This creates an electrically-conducting fluid that is in constant motion."[166] The strength of our planet's magnetic field gives these layers of magma higher viscosity.

But what if the magnetic field of the Earth weakens tremendously, as it has already done over the last few thousand years? Layers under the crust may suddenly shift from high friction layers holding tight against each other to low friction lubricating layers that facilitate sliding. A quote from Wenceslas Jardetsky's article: "Aperiodic pole shift and deformation of the Earth's crust" seems relevant here, despite the fact that magnetohydrodynamics had not been discovered yet at the time of his article and he was only indirectly coming to the relevant conclusions in 1962 when he wrote:

"Other phenomena of this kind should be related to the gliding of the entire crust of the earth and the resulting shear, to the effect of longitudinal displacements... The suggested theories of the pole shift should be modified. It is shown that the tendency of the crust to glide in the way under consideration can be derived from Volterra's investigations of the rotation of a body with internal movements."[167] Authors like Brown, Hapgood, and myself are constantly working to modify the pole shift theory with new evidence and ideas... and not that I understand all the math behind it, but for anyone who is interested there is new evidence from 2017, or at least a new and more detailed explanation describing "the Stokes-Einstein relation (SER) and the Rosenfeld entropy scaling law (ESL) in liquid.... In other words, the SER can be used to calculate the viscosity of liquid Earth's outer core from the self-diffusion coefficients of iron/nickel and the ESL can be used to predict the viscosity and diffusion coefficients of liquid Earth's outer core form its structural properties."[168]

In simple terms: imagine the board and brick from our earlier example, if the brick was on a polished, smooth surface, as if the rough wooden board were stainless steel. The angle of incline needed for it to slide would be much less. Likewise, the crust of the Earth might break free and start a pole shift based on current Antarctic ice cap mass imbalances if the magnetic field weakens enough – and it is steadily weakening already. In a magnetic pole shift, the Earth's magnetic field strength might temporarily fall near zero before new magnetic poles are established. And under the temporary conditions of a very weak magnetic field, the friction between layers below the crust would virtually disappear, and the threshold for a pole shift to occur would be much lower.

Hapgood noted that the crucial layer, the asthenosphere, is usually too thick and viscous to allow the sliding necessary for a crustal displacement (pole shift) with the current forces at play from the imbalance of ice now existing. But it has a more liquid "wave-guide layer" about 100 miles down, discovered by a Soviet geophysicist named Beloussov. Combining a new understanding of magnetohydrodynamics (the idea that viscosity and friction decrease or even disappear in liquids with charged particles if the magnetic field decreases enough) with the newly discovered existence of Beloussov's more liquid layer under the crust, Hapgood said: "At a level where the viscosity of the asthenosphere would be reduced to its lowest point by the fluid wave-guide layer, and so the lithosphere would in effect be borne along a stream flowing in a liquid... friction would be minimized, while viscosity would present no bar to a comparatively rapid displacement... shifts of the lithosphere have at times attained extraordinary speeds... the combination of the geometrical progression of centrifugal effects with the zone of easy shear in the wave guide layer opens up the possibility of extremely rapid movements of the earth's outer shell."[169]

I think magnetohydrodynamics are a crucial component. Liquid metals and liquid rock in a strong magnetic field have a much higher viscosity (more friction) than they do with a weak or non-existent magnetic field. The Earth's magnetic field has been weakening steadily for several thousand years and scientists in early 2018 are starting to admit that a geomagnetic reversal appears to be coming soon. This would mean the magnetic field would presumably weaken even further, possibly going down to nothing temporarily before the new field orientation remerges and strengthens. The viscosity of the layers of magma beneath the crust of the earth would be tremendously reduced.

So while an imbalance of mass above or slightly below the surface is an important destabilizing factor – it is possible that the mass imbalance would never be strong enough to start a pole shift without a weakening magnetic field. Unfortunately, Earth's magnetic field is already getting very weak. Hapgood's dilemma – how could the pole shift be very rapid, yet not move the ice caps more than thirty degrees – could be due to a fast reversal of the magnetic field, with the reestablishment of field strength in a very short period of time (days, or even hours.) Like a magnetic lock that suddenly goes off (and back on again a few days later) the reversing magnetic field of the Earth may suddenly allow a pole shift to start, and may suddenly end the movement a short time later as a new magnetic field emerges – whether the crust of the Earth moved far enough to relieve all the mass imbalances and get the old poles to the equator or not.

I propose that our galactic center has periodic outbursts of energy and that when such a (as Dr. Paul LaViolette puts it) "galactic superwave" hits the Earth its gravitational and electromagnetic components temporarily loosen the crust from the layers below, but that right after this event's wavefront passes by, the equation – the balance of various forces at play – quickly favors stabilization again. The magnetic field strengthens, viscosity comes back, and the pole shift screeches to a halt, perhaps even more suddenly than it began just a brief time earlier. For what it's worth, that is my way of reconciling the evidence for both a fast pole shift and a movement of only thirty degrees latitude…

Hapgood revised his theory for as long as he was alive to adapt and incorporate new information. Perhaps he would be in total agreement with me if he had access to everything I can read and he was still analyzing this in 2018. I wish I had the opportunity to discuss such theories with him. But the real history of the development of Hapgood's ideas started in the early 1950s, and he discussed them with people far more knowledgeable than I am.

Albert Einstein

Hapgood reached out to Albert Einstein for his opinion on the pole shift evidence he was accumulating and putting together into a convincing theory. Einstein responded in a letter to Hapgood in 1953 that said: "I find your arguments very impressive and have the impression that your hypothesis is correct. One can hardly doubt that significant shifts of the crust of the earth have taken place repeatedly and within a short time." Einstein later wrote the foreword to Hapgood's 1958 book: <u>Earth's Shifting Crust</u>, lending great credibility to a theory many would like to dismiss as pseudo-science:

"I frequently receive communications from people who wish to consult me concerning their unpublished ideas. It goes without saying that these ideas are very seldom possessed of scientific validity. The very first communication, however, that I received from Mr. Hapgood electrified me. His idea is original, of great simplicity, and if it continues to prove itself, of great importance to everything that is related to the history of the earth's surface.

A great many empirical data indicate that at each point on the earth's surface that has been carefully studied, many climatic changes have taken place, apparently quite suddenly. This, according to Hapgood, is explicable if the virtually rigid outer crust of the earth undergoes, from time to time, extensive displacement over the viscous, plastic, possibly fluid inner layers. Such displacements may take place as the consequence of comparatively slight forces exerted on the crust, derived from the earth's momentum of rotation, which in turn will tend to alter the axis of rotation of the earth's crust.

In a polar region there is continual deposition of ice, which is not symmetrically distributed about the pole. The earth's rotation acts on these unsymmetrically deposited masses, and produces centrifugal momentum that is transmitted to the rigid crust of the earth. The constantly increasing centrifugal momentum produced in this way will, when it has reached a certain point, produce a movement of the earth's crust over the rest of the earth's body, and this will displace the polar regions toward the equator." Anyone who doubted the possibility of pole shifts might have had to reassess their dismissal of the subject, if they valued the opinions of Albert Einstein.

Einstein's biggest concern with the pole shift theory was not about the evidence that pole shifts have repeatedly occurred – it was over the imbalanced ice cap mass as the trigger. In later years, Hapgood put less and less faith in the idea that the imbalanced ice caps could be the trigger for pole shifts by themselves, and suggested in the updated version of <u>Earth's Shifting Crust – Path of the Pole</u> – that there must be some other trigger in the core of the earth, or maybe even an external, cosmic trigger – but that ice was probably not the sole cause.

Dr. Hapgood continued working on the crust-displacement theory, though his <u>Maps of the Ancient Sea Kings</u> book focused more on the features of ancient maps that showed advanced knowledge of lands and math and map projections that should not have been available at the times they were made. In the Maps book he reviewed what he considered "the first hard evidence that advanced peoples preceded all the peoples now known to history...

ancient voyagers traveled from pole to pole... some ancient people explored the coasts of Antarctica when its coasts were free of ice." And in John White's Pole Shift book from 1980, White said Hapgood was working on a third version of that book with new information and updates that strengthen his theory.

Hapgood wrote in a letter to Rand Flem-Ath (author of When the Sky Fell) in 1982 that "I have convincing evidence of a whole cycle of civilization in America and Antarctica, suggesting advanced levels of science." He may have discovered evidence our government did not want publicity for. I'm not alleging that Hapgood was assassinated or that he was a problem to be gotten rid of at that point, but he was, unfortunately, hit by a car and killed before publishing anything about that... And we will see later that the U.S. Government (and the Vatican) has suppressed and classified some pole shift information, which begs the questions: What do they know? What are they worried about?

Chan Thomas' Classified Adam and Eve Story

I also want to review the pole shift ideas from an electrical engineer and self-proclaimed expert in the field of "cataclysmic geology" named Chan Thomas. My interest in Mr. Thomas begins with a short book he wrote around 1963 called "The Adam and Eve Story." Thomas' basic premise is that catastrophic pole shifts repeatedly devastate the surface of the Earth. He claimed that the last one led to our stories of Noah and the Great Flood, and that the pole shift before that, approximately 11,500 years before 1963 (let's say 9,600 B.C.) led to the stories of Adam and Eve being kicked out of Eden, having to flee a wonderful homeland to survive its destruction and produce offspring somewhere new...

I am not devoting a chapter to Chan Thomas because of the research or conclusions I see in his book. Hugh A. Brown and Charles Hapgood would definitely deserve their own chapters before Chan Thomas, based on their contributions to the pole shift theory. What strikes me most about Chan Thomas' book, and the primary reason to dwell on it for a whole chapter is not because of anything I read in his work. Although many of his ideas make sense to me, they are not revolutionary or new concepts on the topic of pole shifts, or biblical references to them.

My interest is piqued by what I CANNOT read in his book. In September 2017, one of the new pole shift videos on Youtube that I mentioned on my first page alerted me to Chan Thomas and his book, which was CLASSIFIED shortly after it was written. I have only been able to see the version stamped "Declassified in Part – Sanitized Copy Approved for Release" in a pdf file found at the CIA.gov library reading room web site. Let that last sentence sink in.

There were many books describing evidence of pole shifts. Hugh A. Brown published <u>Cataclysms of the Earth</u> in 1967, and had an article in <u>Time Magazine</u> titled "Can the Earth Capsize?" in 1948. Professor Charles Hapgood published <u>Earth's Shifting Crust</u> in 1958 (and the updated <u>Path of the Pole</u> in 1970) and <u>Maps of the Ancient Sea Kings: Evidence of Advanced Civilization in the Ice Age</u> in 1966. They weren't classified. Why was Chan Thomas' book ever classified? Why is it still only partially declassified now?

I can only speculate that Thomas may have:

1) Revealed the correct mechanism or cause of pole shifts
2) Revealed their periodic and recurring nature and correctly timed the next occurrence
3) Revealed plans for surviving the pole shift, including locations to build bunkers
4) Revealed specific evidence of a previous advanced civilization like Atlantis
5) Revealed a way to control and weaponize pole shifts
6) Revealed a course of action for America's enemies to prepare for a pole shift

Cause:

If I am correct, the rotational imbalance caused by 19 quadrillion tons of Antarctic ice (possibly in tandem with subsurface mass imbalances) centered over 300 miles away from the rotational South Pole creates significant tangential centripetal forces that pull on the crust and encourage it to shift north along 96 degrees East Longitude. This is Charles Hapgood's original hypothesis, and an imbalanced ice cap as the cause of pole shifts was also Hugh A. Brown's hypothesis. Einstein's biggest concern with this theory was that the ice cap alone might not be sufficient to overcome friction between layers beneath the crust and trigger a pole shift. I indirectly contacted a pair of scientists through an acronym agency friend many years ago to review my ideas on the math and physics of some associated potential triggers and we were told it would be in our best interests to stop asking such questions. Even without their confirmation, I am quite certain that magnetohydrodynamics and the loss of sub-crustal viscosity in a weakened magnetic field is a crucial component to consider.

Hapgood eventually suggested that the ice cap might play a minor role and that the main imbalance of mass could lie underneath the crust. But it seems that there are significant stresses pushing along this 96 degree meridian, giving the equatorial region at this longitude significant seismic and volcanic activity, just as we would expect if the Antarctic ice cap were the primary source of concern.

There are probably additional imbalances of mass underneath the crust, which are much harder to measure, and may or may not be located on the same meridian of longitude. Convection currents in the liquid mantle could play a role as well. But static friction between layers in the lithosphere is strong and overall, mass imbalances are not yet extreme enough to cause a pole shift on their own. I believe the trigger, the straw that breaks the planet's back, comes from periodic outbursts from the center of the galaxy, as explained by Carl Seyfert and Dr. Paul LaViolette. But since their books were not classified either, causes and triggers are probably not the government's concern.

Periodic Timing:

In online commentary on the classified and only partially revealed nature of Chan Thomas' book, most people suspect that the timing of the next pole shift is probably what the government wanted to keep from the public. Ancient Egyptian priests described their records of many past cataclysms spanning tens of thousands of years to visiting Greek wise men like Solon. Plato described the conversations and quoted the Egyptian high priest Sonchis calling what I believe is the galactic superwave: "the stream from heaven, like a pestilence" and stating that it always returns "after the usual interval." I suspect that the usual interval is half a cycle of precession, around 12,960 years, just like Plato's nuptial number for the marriage of heaven and earth. Diodorus calculated the last pole shift was approximately 12,736 years before 2018 A.D. Many sources indicate we are in the ballpark for the time of the next pole shift to occur very soon.

In the Book of Job, God asks Job if he can bind up constellations or lead them forth to rule over the earth at the appropriate times. As discussed in Douglas Krieger's <u>Signs in the Heavens and on the Earth</u>: "Question: What is the significance of the Mazzaroth spoken of in Job 38:31? Answer: Since the Mazzaroth specifies an interval of time (we know this to be 25,920 years known as the Great Precession of the Equinoxes), any answer to our question that pertains to the Lord asking Job if he can disrupt the time interval..."[170] This implies that the cycle will reach its end on schedule. The Torah, the Old Testament of the Bible, was heavily influenced by Moses' Egyptian education. If very ancient records and traditions were kept for thousands of years by the

Egyptian priesthood, it is quite possible that their knowledge has been carried forward into modern times. "Moses was instructed in all the wisdom of the Egyptians," we are told in Acts 7:22.

Chan Thomas noted on page 24 of his book that Moses may have had access to all of Egypt's ancient records, and suggests that Moses and high priests like Aaron and his lineage may have forwarded the information into the future, eventually understood by the Babylonians who captured the wise men of Jerusalem, like Daniel, and brought them to Babylon. If Egyptian and Jewish high priests, Babylonian astronomers, Persian magi, and Greek thinkers all valued this ancient wisdom, and understood anything about cycles of catastrophic pole shifts – then we should not be surprised to see traces of it in the Bible (and other ancient books.) Jesus Himself spent many years in Egypt – and many people, including the current leader of the Hindu religion, claim He spent years in India too…

It would not require time in India to know the traditions of India, especially when so many were incorporated into the foundations of Judaism. Hindu mythology suggests a periodic cycle of the destruction of the world at regular intervals, in which Brahma is the creator, Shiva is the destroyer, and Vishnu is the preserver. Anyone with knowledge of both Hinduism and Judaism would realize that Judaism started with Hinduism.

Before the founder of the Jewish people, Abraham, came from the east with his half-sister/wife Sarah, their original names were Abram and Sarai. In Genesis 17:5 the Lord said "No longer shall your name be called Abram, but your name shall be Abraham." The original names Abram and Sarai are almost identical to the Hindu high god Brahma and his sister/wife Sarai (or Saraisvati.) In Judaism, Abraham eventually has a son named Isaac with his old wife Sarah. But she first gives Abraham her servant Hagar, to produce another son, Ishmael. In India, the Saraisvati River has a tributary, the Ghaggar. Isaac is from the Sanskrit "Ishakhu" ("friend of Shiva") while Ishmael is from the Sanskrit "Ish-Mahal" (Great Shiva.) Even today, Jews may inadvertently acknowledge the death and destruction attributed to the Hindu god Shiva as they mourn for the dead by "sitting shiva." Biblical stories of Abraham should be understood in the Hindu context of recreating a new start after the death and destruction of the old world.

In the Brahmavaivarta Purana, we are told: "I have seen the world arise and vanish, arise and vanish again, like a tortoise's shell coming out of an infinite ocean and sinking back. I was present at the dawn and the twilight of the Cycles, past counting in their numbers."

2 Peter 3:5-7 reminds us how easy it is to forget such cycles of world destruction:

"It escapes their notice that by the word of God the heavens existed long ago and the earth was formed out of water and by water, through which the world at that time was destroyed, being flooded with water. But by His word the present heavens and earth are being reserved for fire, kept for the Day of Judgment and destruction." Note the phrase "the earth" was synonymous with "the world at that time." If ancient cultures had records from before the last pole shift, and understood anything about how and when and why they occurred, we should expect hints and clues in religious histories like the Bible – no matter how distorted the information may be after many generations passing down oral traditions and translations across languages and millennia.

Chan Thomas suggests (on page 26) that early symbols in the Garden of Eden story should be understood as follows: "the tree (of life) symbolized a mother continent, a parent civilization... [the] serpent represented water, or the ocean; a serpent entwined about the tree signified that the mother continent was surrounded by water... Eve's heel on the serpent's head, showing her victory over the oceans." Cherubim were not angels so much as symbols for legs and foundations... "cherubims of the North, East, South, and West were taken away – meaning that the foundations of the mother continent, in all directions, were removed or destroyed."[171]

Thomas seems to be describing Antarctica's destruction by pole shift. This also seems in line with how God "drove the man out" and "stationed the cherubim and the flaming sword which turned every direction" in Genesis 3:24. Should we understand this to mean that people were driven from their Antarctic homeland by cataclysm, after which God reestablished new cardinal directions and placed a flaming sword turning in the sky (the Aurora Australis) over Antarctica?

Thomas also suggests that the precursor to being kicked out of Eden – eating from the tree of knowledge – should be understood to mean that the people in Eden (Atlantis/Antarctica?) knew from their wisdom and science that a pole shift would destroy their homeland soon, and that they would have to emigrate to safe lands somewhere else. In his multiculturally derived translation of Genesis (in which he incorporates similar stories from the Egyptian Book of the Dead and other source texts from the Maya, from India, etc.) – a version of Genesis tailored to his pole shift theories – Thomas comments on the third chapter of Genesis, retranslating some parts as:

"Adam felt God's call... heard God's warning [but] ...knew not where to find refuge... a warning to leave the land of his ancestors... my ancestors gave me this knowledge... the coming inundations of the oceans has been made known to me... God's warning that though they be descendants of original mankind of the motherland, they should leave it, as it was destined for destruction... by his own toil and sweat he was destined to fight the fight for survival after the inundation."[172] I know of no other sources describing ancient prophecies given to Adam warning him of the last pole shift – this seems to be Chan Thomas' own imaginative interpretation.

In my books, I suggest that the next pole shift is due in the early 21st century. Bible prophecies describe heavenly bodies in the night sky during the end times visions of Daniel, Isaiah, and John. If we compare these descriptions with astronomical software that tells us where all heavenly bodies are at any given time, we can see a correlation to the week of December 21-28, 2019. Most significant to me, this week also sees the sun, moon, and planets move in ways that act out all major steps of an ancient, week-long Jewish wedding ceremony. Can we assume this celestial wedding symbolizes Jesus coming for His bride as the pole shift makes a new heaven and a new earth? I cover this correlation in great detail in <u>End Times and 2019</u>.[173]

After studying Nostradamus prophecies for several decades, I conclude that Nostradamus expects a pole shift at the end of a Third World War between Islam and the West. He repeatedly describes a long conflict between Islamic nations and the Christian West, gradually intensifying over 27 years and culminating in WWIII and immediately followed by a pole shift. In the 1990s, I originally thought this could describe warfare during the years from 1991-2018, with a pole shift in 2019. But it is already 2018 as I edit the book you are reading now, leaving very little time for WWIII to precede a pole shift in 2019... It seems much more likely that Nostradamus' 27 years of increasing hostilities between Islam and the West could start with September 11, 2001 – and end with a pole shift around 2028-2029. I cover these prophecies, and many more describing WWIII, in <u>Nostradamus and the Islamic Invasion of Europe</u>. (Long after writing that book, I learned that Delores Cannon suggests Nostradamus says the pole shift will be in October 2029.)

But none of my books have been classified, so either my pole shift timetable is off, or my readers are so few and my influence so minimal that my thoughts don't matter – or perhaps the government doesn't care about timing the next catastrophe. Or perhaps at this point, we're just so close to the end of the present world that it no longer matters.

Surviving the Pole Shift

If you were certain of what was going to happen, and when it would happen... and you had enough money, manpower, time, and other resources at your disposal, how would you attempt to prepare to survive such a catastrophe? My third best choice would involve building a large number of reinforced ships – rounded metal submarine arks that could withstand huge changes in pressure, and hope that enough of them survive and can congregate at the most desirable new areas for future resettlement. My second best choice would be to select mountainous areas likely to remain above sea level and end up at a warm, temperate latitude – and build networks of underground tunnels. Tunnels hundreds of miles long, with multiple entrances and exits, with several areas which could be sectioned off in case floodwaters or lava or any other problems broke through into any part of the tunnel network. If I were the American government, there would by now be huge and well stocked tunnel systems in the Rockies, the Appalachians, the Ozarks, and the Alaskan Mountains. Certain specific locations are more desirable than others, and Chan Thomas may have pointed out locations that were already being developed or planned for development that no one wanted mentioned.

But my best plan, to develop a location that would not merely increase the odds of survival but guarantee them one hundred per cent, would be to set up an underground base on the moon. On the moon there are no oceans to cause tsunamis, no lava to rise up from below, no desperate hordes of survivors in the aftermath of a pole shift that would only ruin conditions on Earth. Since we had the technology to repeatedly visit the moon since 1969, yet officially stopped going in 1972 and have allegedly never been back... I have to wonder why. Some people believe we were never there in the first place, and that the footage we have seen is all fake. There are some intriguing videos and commentaries on the idea that Stanley Kubrick helped fake the moon landings during the filming of 2001: A Space Odyssey. Entire books have been written on the idea of such a moon landing conspiracy.

Some people believe we encountered extraterrestrials who told us to stay away from their base of operations. Some say we found evidence of an earlier human civilization that made it to the moon before recorded history, and we are working on their artifacts and technology. Another idea is that we have a secret space program kept from the public, and that, regardless of what we found on the moon, we are still secretly going there. If we do have a secret space program, the best place to protect people and knowledge and the diversity of life on Earth is not on Earth during a catastrophe that will wrench the crust from the mantle and devastate the surface of our planet. It is pure speculation, but Thomas may have commented on one or more of these topics that come too close to the truth.

Evidence of Previous Advanced Civilization

Maybe Thomas' work was classified because he commented on a location the government wanted to investigate in secret. Is it possible that mankind has achieved technologically advanced civilization before, only to be destroyed by pole shifts and relegated to barely believable myths and legends? The ancient Greek writer Plato tells us that an empire known as Atlantis dominated the world from "beyond the Pillars of Hercules" (somewhere past the Straits of Gibraltar) in the great world ocean beyond the Mediterranean which the

Greeks called the Atlantic – but when they used the word "Atlantic" it referred to all the world's oceans. Plato described the destruction of Atlantis in a single day by great floods approximately 9,000 years before he wrote about it around 600 B.C. – approximately 11,600 years ago, roughly when evidence from many sciences indicates profound changes in the surface conditions of the Earth ended the Ice Age in North America.

Some evidence indicates that parts of Antarctica were temperate before the last pole shift, and that Atlantis may have been in Western or Lesser Antarctica, with the capital city of Poseidia south of what is now Argentina on the Palmer Peninsula… (Eastern or Greater Antarctica, despite receiving much less snowfall, has by far the thickest ice, as this part of the continent was polar both before and after the last pole shift – possibly even during the one before that.)

The survivors may have established cultural offspring in faraway lands including Egypt. I am reminded of John Anthony West's comments in Serpent in the Sky: The High Wisdom of Ancient Egypt – in which he comments on Egyptian civilization appearing out of nowhere at a high level of sophistication and then declining over many centuries – rather than rising from nothing and gaining sophistication in its early history. "Egyptian civilization was not a development, it was a legacy." S.R.K. Glanville wrote in The Legacy of Egypt: "that the science we see at the dawn of recorded history was not science at its dawn but represents the remnants of some great and as yet untraced civilization."

John Michell, in his epic analysis of numbers and ancient systems of measurement: The Dimensions of Paradise – wrote that "astronomical and other data, encoded in the traditional canon of number, indicate that the level of knowledge in remote antiquity was far higher than in early historical times"[174] or any time prior to the 1970s. Peter Lemesurier, in The Great Pyramid Decoded, suggested the possibility that "civilizations have already existed on earth whose scientific achievements were equal or superior to our own."[175] Perhaps survivors of "the Atlantean civilization who brought back to cataclysm-ravaged regions the skills and graces of civilization, including perhaps the knowledge of how to calculate the time of the next cosmic disaster"[176] are the ones we must thank for ensuring that humanity did not lose everything and end up reduced to the Stone Age.

If West and others correct, survivors from somewhere else brought their culture and technology to Egypt to start over. Egyptians built pyramids we couldn't build today, using millions of stone blocks weighing up to an estimated 468 tons for the heaviest ones. They incorporated advanced mathematics into the design, even revealing accurate measurements like the 365.242 days in a year, a precise value of Pi, the correct size of the Earth, (the Great Pyramid represents the northern hemisphere at 1/43,200 scale) the angle of obliquity of the planetary axis, and the length of a cycle of precession of the equinoxes. The Osirion Temple at Abydos shows granite carvings of what look remarkably like a helicopter, a submarine, and possibly aircraft and spacecraft. Just how advanced might ancient technologies have been?

Could Chan Thomas have pointed out a specific example, and revealed the location of an artifact or commented on a suspected method of achieving anti-gravity that his government did not want to draw attention to?

In ancient India, the Yanta Sarvasva texts include hundreds of pages of details on flying craft called Vimana. Construction techniques, materials, speed, altitude, fuel consumption, control mechanisms, weapons, special food and clothing for pilots, attacking both visible and invisible objects, even the ability to eavesdrop and listen in on conversations in enemy aircraft are topics covered in great detail. And the heated mercury plasma vortex described as an energy source thousands of years ago was experimented on in Nazi Germany and allegedly in the TR-3B anti-gravity craft allegedly built and tested in Area 51 (where I hear Chan Thomas went to work... Remember, he was an electrical engineer who basically disappeared... (I heard a little more about where he worked that I was told not to mention or I'd get in trouble with the government, but it has no direct bearing on pole shifts.)

The Mahabharata texts from the same period of ancient India's early history describes the use of nuclear weapons: "Gurkha, flying a swift and powerful vimana, hurled a single projectile charged with the power of the Universe. An incandescent column of smoke and flame, as bright as ten thousand suns, rose with all its splendor... It was an unknown weapon, an iron thunderbolt, a gigantic messenger of death, which reduced to ashes the entire race of the Vrishnis and the Andhakas. The corpses were so burned as to be unrecognizable. Hair and fingernails fell out."

The Mahabharata also describes an epic war (I cannot tell if it is the same war or not) fought during the transition between world ages. Is this a cosmic war, like the biblical war in heaven, fought during a cosmic upheaval and a pole shift? Or was it a very real war between mighty nations desperately fighting for living space when they had to flee homelands devastated by a pole shift and evacuate submerged or frozen lands quickly? The authors of Hamlet's Mill note: "the epic states unmistakably that this tremendous war was fought during the interval between the Dvapara and the Kali Yuga" age. We find more reasons to associate a real human war with pole shift type events at this moment in time when we note that the Vishnu Parana (4:24) says: "On the same day that Krishna departed from the earth the powerful dark-bodied Kali age descended. The ocean rose, and submerged the whole of Dvaraka."

There are regions where this ancient war allegedly took place, near the border of India and Pakistan, where background radiation levels are still fifty times normal, and the people still suffer unusually high rates of birth defects and cancer... And we can imagine what Russia and China are capable of in the near future, should the North Pole relocate to their border and make their nations uninhabitable. Nostradamus is not the first to suggest a nuclear war immediately by a pole shift; there are similarities in end times Bible prophecies, and the same things may have already happened almost 13,000 years ago.

Aristotle once wrote in Metaphysics that "science has often been developed as far as possible, and then again perished... preserved until the present like relics of the ancient treasure." Michio Kushi, author of Forgotten Worlds, said that based on his analysis of ancient Japanese documents, the world had a global transportation system using advanced boats and airplanes before the Flood, before a pole shift around 12,000 years ago. I know Japan's "Records of Ancient Matters" were written down in the year 712 and they mentioned an original home island called Onogorojima that was "somewhere near the pole" and was an "island homeland near the earth's axis" associated with a central pillar – but I don't know of any Japanese records of ancient high technology. Of course, Kushi may know many things I do not, and Japan may have records of ancient modern transportation vehicles, as Egypt, India, and South America do…

The survivors of a pole shift are too busy struggling to survive the aftermath, and have little time to record how and why things happen – even assuming they understood what had happened. Survivors would also tend to be those away from coastlines and at higher altitudes. Those with primitive, sustainable lifestyles who were in tune with nature and adept at hunting and gathering and farming would have much better odds of surviving after a catastrophe. They may have noticed that the advanced and sophisticated people with books and technology and universities and cities were destroyed. The survivors may have even blamed the acquisition of knowledge as an effort to be like gods which had provoked the gods into great wrath and destruction. Western religions suggest that the original sin was eating from the tree of knowledge and it led to being kicked out of Eden. Instead of trying to salvage and recover technology and knowledge from the remains of the devastated civilization, survivors may purposefully destroy it in a misguided attempt to regain divine favor.

If ancient civilizations did have anything like helicopters, jet fighters, and nuclear weapons – we may not recognize even if it is carved into granite or described in ancient books. Physical evidence might have been destroyed by the cataclysm, the survivors, or the ravages of time. If we are very lucky, our ancestors hid and buried such artifacts for future generations. Of course, if our governments found such things in Egypt, Antarctica, or anywhere else – would they publicize it? I think studying it in secret is much more likely… (Just the kind of thing Chan Thomas might have been assigned to back engineer…)

If the mother civilization was in Antarctica, there could be great discoveries to be found underneath the ice. So I understand the flurry of rumors about Antarctica that became popular by late 2016, when it seemed like too many high ranking officials had just visited or were scheduled to visit Antarctica soon. I spoke to a scientist from the Amundsen Scott research station at the South Pole in 2017, and as I expected, he denied rumors of anything interesting going on in Antarctica. But as I have noted in blog articles (endtimesand2019.wordpress.com) there have been too many high ranking officials there to ignore. I can't prove that ancient advanced technology has been found in ruins of Atlantis under the ice, but:

US Secretary of State John Kerry went to Antarctica on November 8, 2016. Allegedly there to investigate climate change, he went on Election Day, when attention would be distracted by concerns over Donald Trump or Hillary Clinton winning that day's presidential election. Climate change, some scientists are starting to realize, may be due not to mankind's carbon dioxide emissions, but to Earth's changing magnetic field… "A growing number of scientists are starting to worry the magnetic pole shift that seems to be underway, is the real culprit behind climate change. Not man made air pollution, not the Sun, not the underground volcanic activity heating up the oceans, but the slow beginning of a pole shift that has been thought to destroy entire civilizations in the past and be one major factor in mass extinctions."[177]

When Kerry returned he suggested a 35 year ban on private tourism to Antarctica.

Buzz Aldrin was rushed home from Antarctica on December 1, 2016, allegedly for medical reasons. If one is inclined to take conspiracy theories seriously, it may be worth considering that Aldrin walked on the moon, and if there were any interesting relics found there, he would probably be familiar with them. As a 33rd degree Mason, a military officer, and astronaut, Aldrin's trustworthiness and lifetime of experience may uniquely qualify him to evaluate interesting technological finds in Antarctica. Yet he allegedly had a heart attack when he saw something astounding there.

As recently as March 2018 I saw another article with a title asking: "Does melting snow reveal ancient human settlement in Antarctica?" Further down it says: "As the ice sheets of Antarctica continue to melt, experts believe that the lessening snow is revealing a mysterious truth about the South Pole. Satellite images from NASA show a possible human settlement some 2.3 kilometres beneath the icy surface. The images seem to show strange markings in the snow which look manufactured, and as though structures used to stand there. Archeologist, Ashoka Tripathi, of the Department of

Archaeology at the University of Calcutta believes the images are evidence of an ancient human settlement."[178] One speculative article from a British newspaper doesn't prove anything.

But there are widespread rumors all over the internet, including: "A naval officer tells us what he remembers, including seeing a huge opening in the ice in a no-fly area they were crossing with a medical emergency on board. Then he ferried a group of scientists who had disappeared for two weeks and has specifically been warned not to refer again to this subject. As he put it, 'they looked scared.' When they returned to McMurdo, their gear was isolated and they were flown back to Christchurch, New Zealand in a special plane." It sounds interesting... and vague and impossible to verify. These are unconfirmed rumors.

Russian Patriarch Kirill, leader of Russia's Eastern Orthodox Church, had just been to Antarctica a few days after an urgent meeting with Pope Francis – the first meeting between Western and Eastern Church leaders since the year 1054. On December 19, 2016 the Moscow Times reported that Vladimir Putin would visit Antarctica in January 2017. The Russians are also coming back to abandoned Soviet bases in Antarctica and modernizing and expanding them.

The list of high-ranking officials from the United States, Russia, and other nations who made trips to Antarctica around this time is extensive and raises questions. I believe that Atlantis may have been based in Antarctica before being destroyed by a pole shift and frozen over. Ancient maps from before the official discovery of Antarctica (like the Piri Re'is map of 1513) seem to be copied from ancient maps, and show sub-glacial features which have been verified by sonar and radar. It seems very possible that if our governments do find advanced ancient technology or evidence of ancient civilizations destroyed by pole shifts, they would chose not to inform us about the cycle of destruction that might bring another pole shift in our near future.

So I still think that Chan Thomas' book may have been classified for pointing out evidence of past civilizations being destroyed, or for suggesting that future searches should focus on specific sites in Antarctica (or elsewhere) – quite possibly sites that our governments do not want anyone to pay attention to.

Weaponizing Pole Shifts:

What if there were technologies that would allow a nation's army to control or cause a pole shift? Imagine if a country could prepare bunkers for their elite, chose the timing, shut the doors behind themselves and unleash the ultimate disaster to cleanse the rest of the world. As one comment in the new X-Files show in 2016 said, it would be: "the ultimate weapon – the ability to depopulate the planet, to kill everyone but the chosen." (I am reminded of the first of ten suggestions for humanity as carved into the granite monoliths of the Georgia Guidestones. Number One says: "Maintain humanity under 500,000,000 in perpetual balance with nature." This mountaintop monument marking a site on what I expect to be the next equator seems to be a message for the post-pole-shift future, after human population has been vastly reduced from its current levels approaching eight billion.)

If you could control where the next North Pole would be after the catastrophe, you could minimize the destruction of your own homeland, or maximize the destruction of your enemy's lands. For example: Charles Hapgood suggested that pole shifts usually move the poles about thirty degrees, and that the Antarctic ice imbalance has a center of mass along 96 degrees East Longitude. Movement north along this path could put the new North Pole near Lake Baikal, in Siberia, near the border of Russia and China.

If the pole shift does move this way, most of Russia and China will be uninhabitable. Which brings me to another article I wrote in June 2014 and posted on Before It's News: "Third Secret of Fatima – Prophecy of a Pole Shift."[179] Before I quote that article, I must note that in 1967, on the fiftieth anniversary of the events that occurred in Fatima, Portugal in 1917, Pope Paul VI commented on revealing the Third Secret of Fatima and said "terrible damage could be provoked by arbitrary interpretations, not authorized by the teaching of the Church... replacing the theology of the true and great Fathers of the Church with new and peculiar ideologies." That certainly could be a result of acknowledging another pole shift is expected in the near future. This seems even more likely, given later comments from Pope John Paul II in the article below:

"In 1917, the most credible and well-documented visions of the Virgin Mary were received by three young cousins in Fatima, Portugal. Several times that year, always on the thirteenth day of the month, the children experienced trance-like visions in which they claimed the Holy Mother spoke to them. On July 13, 1917, the three cousins were given three secrets – three prophecies about horrible things to come. The first two secrets describe aspects of hell, WWI and WWII, and the importance of Russia as a Christian nation. The Third Secret of Fatima may involve a pole shift and an apocalyptic third world war, but we don't know for certain – because the Vatican refuses to satisfy our curiosity and discuss the details of the Third Secret openly.

By late 1943, two of the children had already died, and Lucia was sick. The Bishop of Leira ordered her to write down the Third Secret, which the Virgin Mary had told her not to reveal. Lucia justified doing so because Catholics are taught that orders from the Church are meant to be viewed as orders from God. Lucia wrote down the Third Secret, with instructions that it should be revealed in 1960, because more would be understood then. But in 1960 Vatican officials wrote in a press release that the information was so horrible that it was "most probable the Secret would remain, forever, under absolute seal."[180] Cardinal Ottaviani related that Pope John XXIII placed the Secret "in one of those archives which are like a very deep, dark well, to the bottom of which papers fall and no one is able to see them anymore."[181]

Cardinal Ratzinger (later Pope Benedict) was always tight lipped about the Fatima prophecies, and Pope John Paul II placed him in charge of them. Ratzinger commented in 1984 that "the prophecies contained in this Third Secret correspond to what Scripture announces." But which scripture – Revelation? [In The Ratzinger Report, Vittorio Messori quotes then Cardinal Ratzinger stating that the Fatima secrets include: "perils that threaten humanity."]

In the year 2000, the Vatican succumbed to pressure and released... something. They claimed the third secret was about the 1981 assassination attempt on Pope John Paul II. Few believed that the first secrets were on big subjects like hell, world wars, and the spread of communism – but that a mere failed assassination was the horrible secret that couldn't be revealed for so long. Bishop Joao Venancio says he saw Lucia's original paper, handwritten with about 25 lines on one page with small margins – and that what the Vatican released in 2000 was 62 lines on four pages with no margins – not the authentic document. We also have interesting comments from Pope John Paul II, who commented on the Third Secret while speaking in Fulda, Germany:

"It should be sufficient for all Christians to know this: if there is a message in which it is written that the oceans will flood whole areas of the earth, and that from one moment to the next millions of people will perish, truly the publication of such a message is no longer something to be so much desired."[182]

Since I assume there is a pole shift coming soon, I am willing to assume the pope knew what he was talking about, and that the Third Secret refers to horrific events at the end of the world. I assume that the official explanation given on May 13, 2000 is a deception.

On March 13, 2013, the Archbishop of Buenos Aires, Cardinal Jorge Mario Bergoglio, became the first pope from the Americas. The Patriarch of Lisbon, Cardinal José Policarpo, soon announced that Pope Francis twice asked him to consecrate his papacy to Our Lady of Fatima, and this was done on May 13, 2013 – on the 96th anniversary of the apparitions of the Virgin Mary to the three shepherd children. Pope Francis obviously takes the secrets of Fatima seriously, and publically acknowledges this by having his papacy consecrated this way. It implies that the events of the third secret are not over.

Fatima author Father Malachi Martin has said that "the Third Secret is your worst nightmare, multiplied exponentially."[183] In an interview with Art Bell in 1997, he explained that "I was shown the Text of the third secret in February 1960. I cooled my heels in the corridor outside the Holy Father's apartments, while my Boss, Cardinal Bea, was inside debating with the Holy Father (Pope John XXIII), and with a other group of other bishops and priests, and two young Portuguese seminarians, who translated the letter, a single page, written in Portuguese..... on the content of the Third Secret, Sister Lucy one day replied: 'It's in the Gospel, and in the Apocalypse, read them.' We even know that one day Lucy indicated Chapters 8, 13 of the Apocalypse.[184] Father Martin said of the fulfillment of the Third Secret that these events are "not 200 years away, it is not 50 years away, it is not 20 years away..." He expressed the view that major portions of end times events would be underway before the centennial of the Marian apparitions (before May 13, 2017.)

Lucia apparently mentioned Revelation chapter 8 in her letter; a chapter that opens the seventh seal. The sixth seal, which sounds like a pole shift, had just been opened in Revelation 6:12-14 "I looked when He broke the sixth seal, and there was a great earthquake; and the sun became black as sackcloth made of hair, and the whole moon became like blood; and the stars of the sky fell to the earth, as a fig tree casts its unripe figs when shaken by a great wind. The sky was split apart like a scroll when it is rolled up, and every mountain and island were moved out of their places."

The seventh seal in Revelation 8:1-9 describes the sun's reaction to the likely cause of it temporarily being black like sackcloth – an incoming cosmic dust cloud that eventually provokes massive flaring, as when sawdust is sprinkled on a fire: "When the Lamb broke the seventh seal... Another angel came and stood at the altar, [the angel at the altar of fire could be Jupiter at the edge of the sun, on Judgment Day, as I described at length in my last book] holding a golden censer; and much incense was given to him... And the smoke of the incense... went up before God out of the angel's hand. Then the angel took the censer and filled it with the fire of the altar, and threw it to the earth; and there followed peals of thunder and sounds and flashes of lightning and an earthquake. And the seven angels who had the seven trumpets prepared themselves to sound them. The first sounded, and there came hail and fire, mixed with blood, and they were thrown to the earth; and a third of the earth was burned up, and a third of the trees were burned up, and all the green grass was burned up. The second angel sounded, and something like a great mountain burning with fire was thrown into the sea; and a third of the sea became blood, and a third of the creatures which were in the sea and had life, died; and a third of the ships were destroyed." This sounds like a massive solar flare or CME (coronal mass ejection) hits the earth when the sun flares up in response to incoming dust and "the light of the sun will be seven times brighter." (Isaiah 30:26)

...Pope John Paul II did talk about massive flooding and pole shift type events killing many people. I believe the next North Pole will be in Siberia, northwest of Lake Baikal, near the borders of Russia, China, and Mongolia. There is a large magnetic anomaly there. The famous Tunguska blast event of 1908 occurred there. The Magnetic North Pole is already moving in that direction. And one of the main founders of the pole shift theory, Professor Charles Hapgood (author of books like Path of the Pole and Earth's Shifting Crust) also felt this was the likely location of our next North Pole. If that is correct, and if that will happen soon – what will that do to the people of Russia and China?

Most of their nations would become uninhabitable, arctic wasteland. We may already think of Siberia that way, but imagine it even closer to the North Pole. Let's now look in context at what Pope John Paul II said right before commenting on huge floods killing millions: "Given the seriousness of the contents, my predecessors in the Petrine office diplomatically preferred to postpone publication so as not to encourage the world power of Communism to make certain moves. On the other hand, it should be sufficient for all Christians to know this: if there is a message in which it is written that the oceans will flood whole areas of the earth..."

Pope John Paul II suggested that World Communism – the major powers of which were Russia and China – would make some hostile moves if they knew the secret. Father Malachi Martin agreed and wrote in his book <u>The Keys of This Blood</u>: "were the leaders of the Leninist Party-State to know these words, they would in all probability decide to undertake certain territorial and militaristic moves against which the West could have few if any means of resisting." I think they would decide to conquer new lands farther from the upcoming new North Pole to provide suitable new homelands for their populations. Central Europe, South Asia, and the Middle East would probably be invaded by Russia and China. This may be one reason the Russians are preparing to invade Ukraine. Perhaps in Revelation 9:16 when an army of two hundred million comes to the Euphrates from the east – it is because the imminent pole shift has become obvious and the people of China and Russia are fleeing. If so we can expect huge wars in Europe, Asia, and the Middle East soon."

My article from 2014, quoted at length above, left out commentary on one additional interesting point: the Tunguska Blast of 1908. On June 30, 1908 there was a huge explosion in the sky above Tunguska, Russia – a part of Siberia not far from Lake Baikal. With an explosive force estimated between 10 and 15 megatons, trees were flattened over thousands of square kilometers. The most common explanation (and I admit, the most plausible one) is that a small asteroid or huge meteor or small comet from the Taurid meteor showers vaporized in the atmosphere just before crashing to Earth. There is no blast crater. But some believe that the Serbian-American inventor Nikola Tesla was responsible.

Tesla had very public disagreements with Thomas Edison about the transmission of electricity. In the late 19th century these two geniuses waged a "War of Currents" over the use of direct current (DC) or alternating current (AC) but Edison won, and Edison General Electric, now known as General Electric (GE) remains one of the most successful corporations in the world. General Electric was in business to make money, and did a very good job of it, providing and selling electricity for profit.

Tesla was obsessed with the wireless transmission of energy, and the idea of providing nearly infinite and free electricity to the world. He planned a transmission tower for a world system that used the entire planet as a giant dynamo. Built on Long Island, NY, the Tesla Tower was 187 feet high. Tesla claimed he could produce tens of billions of watts and that if released in "'an incomparably small interval of time,' the energy would be equal to the explosion of millions of tons of TNT, that is, a multi-megaton explosion. Such a transmitter would be capable of projecting the force of a nuclear warhead by radio. Any location in the world could be vaporized."[185]

It has been suggested that Tesla was aware of the competing attempts by Frederick Cook and Robert Peary to be the first men to the North Pole. Both eventually returned to civilization and both claimed to have reached the North Pole in April 1908. But for many months one could only assume that at least one of the expeditions was somewhere near the pole. Some suggest that Tesla decided to send a huge burst of electricity to the North Pole on June 30, hoping that his directed energy would not do any damage to a populated area yet would be seen and reported by the explorers near the North Pole. It has been suggested that Tesla overshot the North Pole and caused the Tunguska blast. I assume that blast was probably due to a Taurid meteor impact. The odds seem much lower that Nikola Tesla's energy beam was the cause, but on the off chance he was the cause, I suggest another possibility, outrageous as it may seem: perhaps that much electromagnetic energy had an unexpected interaction with space-time, and the energy burst did come down at the North Pole – the next North Pole, where it will be in the future.

We will probably never know. Tesla would have had little reason to claim responsibility for the disaster if he still wanted to promote the free wireless delivery of electricity. He had an uphill battle against a powerful company making tons of money metering and selling electricity across transmission wires. And if he did aim for the North Pole and hit Siberia instead, he may have had no idea why his aim was so inaccurate, and may have wanted to correct that problem before acknowledging a huge failure.

Tesla kept working on inventions until his death in 1943, shortly after claiming that he had perfected a "death beam" weapon. America was fighting WWII and took no chances that his ideas for advanced weapons would fall into Nazi hands, or even Yugoslavian hands that might eventually reach Moscow... Sava Kosanovic, Tesla's nephew, a Yugoslav diplomat, and a suspected communist party member – reached his uncle's room the morning after he died, only to find that the body was already removed and that the FBI and other government agencies had already taken everything they wanted from his room, including numerous notebooks and technical equipment. The National Defense Research Committee of the Office of Scientific Research and Development later admitted that Tesla's final years had been spent focused on the wireless transmission of energy and beam weapons. Tesla's papers were allegedly sent to Patterson Air Force Base and led to "Project Nick."

The FBI finally released "all" of the Nikola Tesla documents 73 years later, in 2016. Although some of Tesla's papers were eventually released, many believe the majority of the most interesting ones were so far ahead of their time that they are still under lock and key.

What, if anything, does this have to do with the possible weaponization of pole shifts and the still classified writings of Chan Thomas?

1) Tesla's wireless transmission of electricity may have flattened the future North Pole.
2) Tesla continued working on beam weapons, death beams, and wireless transmission of energy.
3) Chan Thomas, after writing <u>The Adam and Eve Story</u>, was hired by a major defense contractor to investigate reported UFO landing sites and record data on electromagnetic anomalies.
4) Thomas was then hired to develop electromagnetic weapons for the same company.
5) Thomas spent the rest of his career at the Skunkworks – Area 51.

At least, this is what I'm told via sources at agencies with acronyms that have far greater access to information than I do. Allegedly, his disappearance from official records seems so complete and extreme, even to those who can usually track someone's career, Area 51 or not… that they found his absence from lists after a certain date quite intriguing.

I tried to verify this myself and found one article written by MUFON director Dr. Robert Wood stating that in 1966 he worked with a team at McDonnell Douglas studying UFOs, and that he hired Chan Thomas onto his team less for his electrical engineering background and more for his writing on "cataclysmic geology" and because he seemed to have psychic abilities that allowed some level of occasional contact with extraterrestrials!

Wood liked that Chan Thomas was "especially innovative" but also "had such outlandish claims that even some in our own group had difficulty accepting him." They worked with the American Institute of Aeronautics and Astronautics at Vandenburg Air Force Base on both Research & Development and on "getting new business" by attempting to predict the future development of space travel and analyzing UFOs was a major part of their research. Wood also wrote: "Once, when I let Chan talk to an actual customer, he embarrassed us terribly when the guy called up and asked me why I hadn't fired him long ago. His value to me was his tremendously innovative mind. He was a total 'out of the box' thinker."[186]

Did Thomas eventually work on secret projects developing electromagnetic beam weapons? It seems likely. Even if true, this doesn't necessarily mean that a weapon could be used to cause a pole shift or to direct the path of the planet's moving crust so as to choose the location of the future poles... But if this were possible, it would certainly be an effective deterrent against an otherwise superior opponent. Imagine a scenario in which Russia and China, acting together, could win a limited nuclear war against the United States – but they also knew that if they start such a war, America would initiate a pole shift in response, one that would leave most of Russia and China frozen wastelands in the new Arctic... Imagine if any of those nations could steer enough electromagnetic energy to affect the location of magnetic North, and if this is crucial in determining the location of the axis of rotation as well. Pure speculation, I admit – but not outside realm of possibility for an electrical engineer at the Skunkworks with a huge interest in pole shifts. If Thomas had written about ideas on this topic in his book, it could explain the secret classification.

(This reminds me of Cleve Cartmill's story called "Deadline." In 1943 Cartmill was writing for the magazine <u>Amazing Science Fiction</u>. He told the editor, John Campbell, that he had an idea for a story on the Allies developing a futuristic superbomb to use against the Axis. He started researching current scientific articles and decided that refining Uranium-235 could lead to a nuclear fission bomb of such remarkable destructive power that his scientist characters would have long arguments on the moral dilemmas of using their weapon. Cartmill and Campbell also noticed that many scientists who subscribed to their magazine had moved to Los Alamos New Mexico recently. The story got uncomfortably close to reality, prompting an FBI investigation into Cartmill, Campbell, and other prominent writers for the magazine like Isaac Asimov and Robert Heinlein. Cartmill was soon told not to write anything about military uses for uranium for the remainder of the war.)

Preparing America's enemies for a pole shift:

The last of my six thoughts on why Chan Thomas' book might have been classified is that he may have made a suggestion that could help Russia and or China prepare to dominate the post-pole shift conditions in what could effectively be called the next world age. If those nations knew what is coming – if a pole shift is about to make most of their territory uninhabitable – then American leadership may have felt the same way as Pope John Paul II said the Vatican did when they: "diplomatically preferred to postpone publication [of the Third Secret of Fatima] so as not to encourage the world power of Communism to make certain moves." Or as Malachi Martin suggested, "were the leaders of the Leninist Party-State to know these words, they would in all probability decide to undertake certain territorial and militaristic moves against which the West could have few if any means of resisting." If Chan Thomas had suggested the probable future location of the North Pole near where Hapgood suggests it will end up – near the border of Russia and China – he might have suggested a variety of logical steps those nations could take to avoid annihilation and take control of the rest of Europe and Asia before the next pole shift.

It could involve anything from conquering living space from other nations to using technology to change the course of the pole shift. Perhaps Thomas' pole shift book may have been classified because he touched on a variety of things falling under the vague heading of "National Security Interests."

* * * * *

The first section of Thomas' The Adam and Eve Story is called "The Next Cataclysm." He describes the destruction of cities and continents by raging winds and tsunamis up to two miles high. Most areas he describes are flooded, then frozen. Of course there are two pivot points on the earth where latitude and elevation see little change; he suggests that the south half of Africa will fare well during the next pole shift. He suggests that the North Pole, instead of stopping after a mere thirty degrees as Hapgood suggests, will continue almost entirely to the equator and come to rest near Burma. The ice caps in Greenland and Antarctica, he suggests, will completely melt within 25 years, raising sea levels 200 feet from wherever they settle after the pole shift screeches to a halt in just six days of movement, as the book of Genesis describes the last (re)creation of the world. Thomas suggests most of the movement will occur in a single day, "with the poles moving almost to the equator in a fraction of a day" and that there will be very few survivors: "A new era! Yes, the cataclysm has done its work well. The greatest population regulator of all does once more for man what he refuses to do for himself, and drives the pitiful few who survive into a new stone age."[187]

Thomas wrote that he felt challenged by those who had written about great cataclysms to find the cause of the universal destruction which left the evidence they described. "The challenge wouldn't leave me alone. Like a hunger, it gnawed at my subconscious. I could hear the deep tones of Cuvier's challenge, 'find the cause of these events...' I felt Hibben's challenge later on, prodding: '...answer all of the facts.' I decided that this cataclysmic concept, this catastrophic end which seems to visit our planet time after time, needed verification."[188]

This certainly seems to imply that Thomas felt he found the answer!

He soon wrote of comparing the evidence from different scientific backgrounds: "...between-science correlation showed indeed that the concept was true. Not only did it verify that the events have happened, but disclosed when the last five cataclysms were, and what positions the shell of the Earth has been in for the last 35,000 years."[189]

This is informative! Thomas clearly says "the last one, 6,500 years ago, was Noah's Flood!" Perhaps the previous four are spaced out approximately 6,500-7,000 years apart, in order to fit in the 35,000 year window? This would correspond with Hugh A. Brown's suggestion of an approximately 7,000 year cycle... and with the somewhat popular notion that the major pole shift happens every precession cycle of about 25,800 years – with a secondary level catastrophe halfway through after about 12,900 years and lesser events every quarter-precession cycle, after about 6,450 years. Of course, this would mean that – at least approximately – we are due for a relatively major pole shift very soon.

Peter Gutilla wrote an article in <u>Saga Magazine</u> in 1970 about Chan Thomas and his theories called "The End of the World." He asked Thomas when the next cataclysm will occur, to which Thomas replied: "We must face the fact that a cataclysm is a normal part of the earth's life cycle... Sometime between 30 and 500 years from now... There are signs indicating the approach of a magnetic null zone at a rapidly accelerated rate... With adequate funding we could mathematically tie down the exact time and make suitable preparations for it... When it does happen, we will have an Adam and Eve story similar to that of 11,500 years ago, and a Noah story similar to that of 6,500 years ago also. The survivors will be driven into another Stone Age..."

I find it interesting that Thomas felt the evidence pointed to a pole shift in the first half of our current millennium, but far more interesting that he felt properly funded science could calculate the exact timing.

If you have read my earlier book, End Times and 2019 – you know this doesn't surprise me at all. Like Isaac Newton, I believe that with enough information, we can solve the cosmic cryptogram and determine the timing of such an event. Everyone knows the Mayan Long Count calendar ended on December 21, 2012. What most people don't know is that many Mayan leaders say the real date to worry about may be seven years later, and that only Europeans and Americans were focused on 2012. The Dresden Codex of the Mayan creation story, The Popul Vuh, has a drawing depicting a planetary alignment at the end of the world in the next great flood: Venus, Mars, Mercury, and Jupiter are all together at the Milky Way. The alignment is common enough that several future dates are a match. But December 2012 was not. December 2019 is a match.

Did the Maya foresee a period of seven years of turbulence at the start of the new cycle, much like the Christian idea of a seven year tribulation at the end of the old cycle? For what it's worth, Graham Hancock – a brilliant researcher and author of books like The Message of the Sphinx, Fingerprints of the Gods, Heaven's Mirror, Lost Knowledge of the Ancients, etc.… said on a recent interview on The Joe Rogan Experience that because the Mayan calendar is largely based on the position of the winter solstice sun between Scorpio and Sagittarius near the dark rift at the center of our Milky Way galaxy – as It Is right now for several decades during this brief window of years (approximately 1980-2018) in the precession cycle – "the story of the Mayan calendar isn't quite over yet."

I think many cultures see the same future events... The Chippewa medicine man Sun Bear once said "It is sealed in the spirit world" when telling John

White what "he saw in the ancient prophecies of various North American tribes that foretell a time of terrible tribulation."[190] Without focusing too much on the time I concluded many prophecies point to (December 2019) we can at least note that John Major Jenkins wrote in Maya Cosmogenesis that the Pyramid of Kukulkan at Chitchen Itza "is a precessional alarm clock set for the twenty-first century."

The Great Pyramid in Egypt may do the very same thing. It is believed to be a representation of the Northern hemisphere of the Earth at 1/43,200 scale, representing the entire planet if we assume an equally sized portion below ground. Many believe that the missing capstone which was never on top of the pyramid represents that the world is not complete. Some view this as spiritually symbolic of man aspiring towards godliness. Others suggest that if completed up to a pointy top, the Great Pyramid would be 5780 inches high, that inches are not just an English measurement but an ancient one, and that 5780 in the Hebrew Calendar corresponds to a potential completion of the world in 2019 or 2020.

Scott Creighton and Gary Osborn tackle this timing from a completely different "angle" in The Giza Prophecy. They believe that angles between key points on a circle drawn around the Sphinx and the pyramids leads to a "Giza-Orion time line" with past and future dates indicating cosmic destruction. Their next future date range they are concerned with is right now, approximately 2000-2025 A.D. Creighton and Osborn conclude that "the Giza pyramids... served as an astronomical clock or calendar that would allow their descendants to know the precise timetable of this deadly Earth calamity and perhaps also the next date of the cycle."[191]

Does Chan Thomas expect a pole shift in the early 21st century too? I have explained at length that I wondered what Chan Thomas discovered that led to his book being classified. He gives us a hint when he wrote: "The challenge was really two-fold: Find the process – what happens in a cataclysm; and the trigger – what causes a cataclysm to start. What a chase! And what a dramatic story of the earth's history we uncovered: Civilizations of 20,000 years ago more advanced than our wildest imagination..."[192]

There's nothing about that in the declassified version of his book! I have to assume he initially said something about this topic, and it was classified...

As for processes and causes... as other pole shift investigators before him had also noticed, Thomas realized that one of the molten layers beneath the outer crust of the earth is normally thick and viscous, acting like an almost solid plastic glue holding the crust in place over the core. But the earth's magnetic field is what helps keep this layer like a solid plastic. Take the electromagnetic field away, and this layer loses viscosity and friction and the crust will decouple from the core. Then quite suddenly, the tangential centripetal forces of the mass imbalance near the South Pole will "correct itself" by moving toward the equator. We can witness this effect experimentally in a container of liquid mercury with a float on top. Without an electrical field present, we can spin a blender-like agitator at the bottom of the container, and no spin is transferred up to an object floating high above on the surface of the mercury. Add an electrical field, and the mercury becomes more plastic and the float will spin. Remove the field and the float decouples and stops spinning.

In his Adam and Eve Story Thomas wrote: "Every few thousand years the magnetic and electrical orderliness inside the Earth is disrupted, and the molten layer is allowed to act like a free liquid, which it was all the time anyway. It then serves as a lubricant for the ice caps to pull the shell of the earth around the inside... and all hell breaks loose. The atmosphere and the oceans don't shift with the shell – they just keep rotating West to East."[193] I doubt that Thomas has correctly assumed that the pole shift is complete within hours – but based on this he expects thousand mile per hour winds and two mile high tsunamis flooding over continents. I disagree for a few reasons, including that human civilization has survived these events many times before, and also because magnetic evidence in lava rock and the Book of Genesis in the Bible suggest an event that lasts six days. This would still lead to intense storms and great floods; but there is a big difference between exterminating 99.9999 per cent of all life on the surface of the Earth in a few hours, and exterminating 80-90 per cent over a few days.

"Now what about the trigger?" Thomas asks, "It had to be something natural, a part of nature's ordinary structure, which disrupts the Earth's inner electrical and magnetic structures whenever it happens. It also has to be a kind of happening which decreases the inner electrical and magnetic forces to the extent that they cannot support keeping the shallow molten layer acting as if it were plastic..." He briefly discusses magnetohydrodynamics and the increased viscosity a strong magnetic field gives to liquid magma, and how a cosmic ebb and flow of energy seems to periodically be responsible for "diminishing the Earth's inner mhd energy to so low a level that the shallow molten layer, starting at 60 miles deep and extending to 120 miles deep, is allowed to act as a free liquid lubricating layer between the earth's shell and solid interior."[194]

As for the trigger that causes the change in the electromagnetic field of the Earth, Thomas considered the role of changes in the sun (the focus of Dr. Robert Schoch in his book, Forgotten Civilization) and what would happen when "a 'dead' galaxy explodes" (somewhat reminiscent of the galactic superwave theory I take most seriously, as detailed by Dr. Paul LaViolette in Earth Under Fire.) Thomas wrote of "concentric spheres of magnetic energy that fill the galaxy, presumably from the center of the galaxy. Their effect on the earth's magnetohydrodynamic energy in turn makes the magma layer a lubricant for the solid crust, allowing it to slip around as it is pulled by the mass of the ice cap."[195]

YES! It is wonderful to find someone new in almost complete agreement with my own theories and with Dr. LaViolette's "galactic superwave" theory! But then Thomas dashes my euphoria by downplaying the cosmic role and stating a preference for a different trigger altogether... Thomas settles on changes within the core of the Earth itself, with "neutral matter" occasionally escaping the core and disrupting the electromagnetic current of the planet.

I do not agree with his conclusion on this unexplained eruption of "neutral matter"... but I do agree with what he says next: "Ice ages are not a matter of advancing and retreating ice; it's simply that different areas of the Earth are in polar regions at different times." Thomas lists five different previous North Pole locations, each lasting somewhere around 5,000-10,000 years.[196] I do not agree with all of his conclusions. For example I do not think that Egypt could have been in the arctic "recently." Like Hugh Brown, Thomas suggests that the Sudan Basin was the site of the North Pole between our current Arctic Ocean North Pole and the one I (and Charles Hapgood) believe came before it, with the North Pole in Hudson Bay, which Thomas lists as the location two pole shifts back, instead of one pole shift back like Professor Hapgood does...

The pdf file in which the CIA "partially declassified" Chan Thomas' book would have more to it if it weren't merely partially sanitized for public view. I have read the declassified sections repeatedly and still can only speculate that Thomas believes Antarctica was Atlantis and Eden, and that he must have written something specific – I know not what – that our secret services do not want publicized. When I first saw this pdf file on the CIA web site in September 2017, it had several pages of unrelated magazine articles at the beginning, and a receipt and a list of supplies for someone's household repair project at the end. I wondered if the unrelated pages were there to compensate for missing censored book pages, to make the whole pdf file appear to be the right length and match the page numbers of the book.

But one insightful remark online suggested that the CIA is commenting in a very subtle way on the pole shift topic with these seemingly unrelated pages. The magazine articles from the following year show that even after the world was warned, no one cared and it changed nothing. The unrelated articles at the beginning show the everyday topics the public focuses on instead of paying attention to pole shifts. The "supplies" list at the end hints that you should be prepared and have supplies for when the next pole shift happens.

But the receipt for the supplies list that was there in September 2017 was no longer at the end of the file on the CIA's pdf in early 2018 – it had been removed. It wasn't a list of the supplies you would want after a pole shift; it seemed irrelevant. But its removal from the file might suggest a hint that time is running out to buy what you need to be prepared.

In April 2018, I was reviewing the file again and a supplies list is back – but it isn't the supplies receipt I remember, and now it's a substitute for whatever Thomas had written on page 48. Page 48 was undoubtedly missing all along. I would consider the possibility that my memory is faulty and it was always that way – but I spoke to someone else who remembers seeing the different version last fall, and agreed the receipt was missing a few months ago, so I'm assuming my memory is correct and it has changed. Why?

I may be reading too much into the list of items on the receipt, but with surviving pole shifts and flooding and a diminishing magnetic field in mind, I can't help but notice:

A "magnetic key holder" – does magnetism hold the key?

There are three references to lights – a blinking warning light, a map reading light, and a Monoway handy light... Could this be a "warning" that we are going to need to read a "map" for our one way (Mono-way) flight to safety?

There are two references to water – a sprinkler, and a car washing brush... Will water be a problem to prepare for?

There are two references to drilling holes – an auger bit and a hole saw set... Are we getting a hint about tunneling into our bunkers?

There are two references to electric batteries – C Cells and a 9 volt battery charger... Is the Earth going to get an electromagnetic recharge?

Maybe I'm grasping at straws... or maybe the C.I.A. chose that list for a reason. I can't stop myself from speculation. When the C.I.A. does something it isn't arbitrary. They classified Thomas' book for a reason. They present it now, just the way they chose to do so, for a reason. I would love to understand their thought process...

Space Age Evidence, Modern Theories, and Ancient Legends

As our technology improves we can look deeper into the Earth and deeper into space. Some of what we find in both directions indicates cycles of mass destruction by pole shifts. I wrote an article about some of the findings which was used by FATE Magazine in February 2015 called "The History and Truth About Pole Shifts on Earth." Right now the link seems to have been removed from access by search engines (I wonder why) but the content can still be found through Google's webcache option. I am including about half of the article below, with a few new comments edited in for clarification.

"Look up 'pole shift' online and you will find many opinions ranging from 'they are rare magnetic events' to 'civilization will be destroyed soon and the evidence is being suppressed!' Some believe there are merely magnetic pole shifts, in which only the magnetic field rotates far away from the North and South Poles we know now. In a purely magnetic pole shift, relatively little would be affected. The sun would still rise in the east and the surface of the earth would remain in the same position. Rotational pole shifts would be more impressive – in which the axis of rotation changes and the geographical poles change location. There is also great debate over how long a pole shift takes. No one would notice if this merely took the form of continental drift over millions of years. But some believe a major shift could occur in days, or even hours. Such a rapid geographical pole shift would destroy civilization – and that is the type of pole shift covered in this article, because much evidence suggests this is the type of pole shift that regularly occurs.

Some theories assume that the planet will move as a solid ball, with the various layers of the crust, mantle, and core all remaining in place relative to each other – but not in relation to the heavenly bodies in outer space. The planets, moon, and sun would appear to us to be wandering off course while the earth remains still. World mythology has many legends of the sun-god's chariot veering out of control for a reason. The new path of the sun's apparent travels would redefine our notions of east and west, and north and south. Ice would start accumulating at the new poles, and melt at the old ones. Changes in sea level, latitude, ocean currents, and wind patterns would radically alter our weather.

…Imagine that the surface layers of the Earth's crust are not permanently attached to the inner core, and that the crust periodically moves over the core as one solid piece like the outside of a chocolate covered cherry. If the core of the earth maintains its rotation while the surface or crust of the earth detaches and moves on its own, then we would keep our 23 degree tilt and our seasons – but in addition to many of the huge changes described above – we would also experience enormous earthquakes, tidal waves, and volcanic activity. Just the momentum of entire oceans of water crashing inland for hundreds of miles could exterminate billions of people, and countless species of animals and plants in a single day. Unfortunately, most evidence indicates this is the form of pole shift we need to worry about.

Huge pressures often build up below the Earth's crust, especially under the weight of polar ice caps. Just as earthquakes frequently correct a minor, localized adjustment of imbalances – a crustal displacement (rotational pole shift of the earth's surface) is the process for a major, global correction. This type of pole shift – a movement of the entire crust of the earth over the molten interior – is not to be misunderstood as merely a magnetic pole shift. The idea that the magnetic polarity of the planet can flip or reverse or establish new locations for magnetic north and south is widely accepted. And it should be – magnetic north has been moving at an accelerating rate in recent years. Movement of 40 miles per year towards Russia was documented up to about 2010, and since then government data on the presumably accelerating magnetic pole shift has been removed from web sites the public can access.

This is no surprise. Professor Hapgood noticed the accelerating movement of the Magnetic North Pole back in the 1960s, and commented on it both in <u>Path of the Pole</u>, and even more in an interview with <u>Argosy Magazine</u> in October 1973, for the article "Shifting Continents" when he said that "if it is accelerating at the rate suggested, then it is quite possible that within a century a state of chaos may embrace civilization,"[197] because a shift in the magnetic poles probably leads to a pole shift of the axis of rotation for the surface of the earth. Hapgood warned: "There is only one hope of salvaging civilization from a total wreck, a wreck that has perhaps occurred several times in the past, and this is through some form of centralized world government that could organize the shifting of populations..." (Of course, this salvation of the human population would be much easier if population levels were vastly reduced first and only a small fraction of humanity is selected to survive...)

As suggested above, magnetic pole shifts may just be a precursor to a crustal displacement, when the surface of the earth rotates as one solid piece over the molten core below. This is not continental drift, with continents moving slowly apart over hundreds of millions of years. A rotational pole shift involves all continents moving together in unison, covering a vast distance over the core below in days or hours.

When it happens, the bulk of the Earth's core will keep rotating west to east just as it had been underneath us, but the surface we live on will suddenly start moving in a new direction that rotates the extra mass of the Antarctic ice cap towards the equator. Under normal conditions, the friction between layers in the earth's crust is enough to prevent one layer from sliding over another. But mass imbalances and a weakening magnetic field contribute to layers in the crust reaching their 'plastic limit' and liquefying suddenly and completely, just like when a tree branch finally breaks under pressure.

Two locations at opposite ends of the earth will act as pivot points. Near those relatively safe locations, (as Egypt was fortunate to be somewhat near a pivot point last time) latitude and altitude changes will be minimal. Few species will face extinction there; and human civilizations will probably survive there. But locations farther from the pivot points will suffer great and drastic changes. What we now consider the North and South Poles will move towards the new equator. Old ice caps at new latitudes will begin to melt rapidly, while some equatorial lands will move towards the new Arctic and Antarctic and start an eternal snowfall on lands that may have been 50 to 100 degrees warmer the day before.

Consider the famous 'Beresovka Mammoth' that was flash frozen thousands of years ago with about fifty pounds of vegetation including seeds, beans, and summer flowers unchewed in its mouth and undigested in its stomach. Imagine how much body heat a mammoth must have while alive, and how quickly chewed vegetation in stomach acid would deteriorate if not quickly frozen. What kind of temperature drop must have occurred to freeze a mammoth solid so quickly?

Temperatures will not be the only change that accompanies the shift in latitude. The Earth has an equatorial bulge, and most of it may be seawater. Lands moving away from the equator will tend to rise higher as they depart this bulge of seawater, and lands approaching the equator may quickly submerge below sea level – like the "myth" of Atlantis sinking beneath the waves. Around the world, stories like Noah's flood are told by many cultures for a reason. (And though I assume pole shifts usually cause such destruction, a very credible theory for the cause of Noah's flood is a comet impact in the Indian Ocean about 5,000 years ago.)

Off-the-charts earthquakes, tsunamis, and volcanoes will also bring death to entire regions long before changes in temperature. And for perhaps the first time, the people who do survive the natural disasters will also be poisoned by all of the world's damaged oil refineries, chemical plants, and nuclear reactors. Assume that no government or military will be left intact to help clean up or rebuild. America, and great achievements like landing on the moon, could be reduced to myths and legends.

Hugh A. Brown and Charles Hapgood attributed the cause of pole shifts to the imbalances from ice that is not centered at the rotational poles. Antarctica's ice cap is about 19 quadrillion tons, and is centered over 300 miles from the pole. This causes a tangential centripetal force which pushes the ice cap away from the pole and towards the equator. Similar issues are known to have caused pole shifts on other planets and moons.

The biggest gravity anomaly in the solar system is the Tharsis volcano on Mars. Now sitting on the Martian equator, it is believed to have migrated there due to rotation torque on its extra mass. Just like high objects fall down when given an opportunity to reach a lower potential energy state, a rotating planet will "fall" to a slightly lower-energy axis if too much mass accumulates far from its equator. Jupiter's moon Europa also shows evidence of a pole shift of almost 90 degrees, and scientists tell us the cause was probably due to an accumulation of ice at the former poles.

As Nancy Atkinson wrote in her article: "Pole Shift on Europa" in Universe Today Magazine in 2008: "Curved features on Jupiter's moon Europa may indicate that its poles have wandered by almost 90 degrees, a new study reports. Researchers believe the drastic shift in Europa's rotational axis was likely a result of the build-up of thick ice at the poles." If the mass imbalance of a huge volcano or ice cap can cause pole shifts on Mars and Europa, why not on Earth?

Even Albert Einstein endorsed these concepts. In a forward to Hapgood's book 'Earth's Shifting Crust,' Einstein wrote that 'In a polar region there is continual deposition of ice, which is not symmetrically distributed about the pole… The constantly increasing centrifugal momentum produced this way will, when it reaches a certain point, produce a movement of the earth's crust over the rest of the earth's body, and this will displace the polar regions towards the equator.' Einstein also said 'One can hardly doubt that significant shifts of the crust of the earth have taken place repeatedly and within a short time.'

There are comments on this at poleshift.com: 'This concept is not accepted by orthodox science, although it has had some distinguished fans, including Albert Einstein. While there is evidence for displacements in relatively recent geological times, the sticking point is a lack of internal mechanism.' The author of that site agrees that even Einstein said the pole shift idea makes sense, but that mainstream science does not accept or promote the theory because the imbalance of ice is not viewed by most as a sufficient cause – leaving nothing on earth to clearly explain why a pole shift happens.

Other scientists suggest additional trigger mechanisms may be necessary which are not internal – not on earth – because the ice caps may not (yet) be large enough to initiate a pole shift on their own. Dr. Robert Schoch's 'Forgotten Civilization' emphasizes the role of periodic solar outbursts in destroying a previous civilization…" [From this point on I am no longer quoting my old Pole Shift article from FATE Magazine word for word, but am editing in some new material.] stating that "our sun is not as stable as most people generally believe." In Schoch's book he acknowledges Dr. LaViolette's conclusions, writing that "LaViolette has argued that a solar proton event (SPE) hit Earth at the end of the last ice age. The galactic superwave may have initiated the solar instability." Dr. LaViolette tells us: "a mass extinction episode... appears to have been brought about by a solar event, possibly a nova-like outburst, triggered by the incursion of cosmic dust."[198] Henry Kroll, commenting on periodic extinction events in <u>Cosmological Ice Ages</u>, said: "The Earth's biosphere is not a closed system... The fact that the extinctions are regular implies... galactic pulses coming from outside the solar system to affect the history of life on earth."[199]

Evidence from ice core samples shows that at the end of "the last ice age, dust was accumulating on the Earth's surface hundreds of times faster than it does today."[200] High acidity and concentrations of iridium, nickel, tin, gold, platinum, and other metals known to be found in higher concentrations in meteorites than in the surface of the earth, along with dust grain sizes too large to be carried far by wind, support the cosmic origin of the dust. (Many people assume that asteroid impacts caused these spikes in various metal concentrations but impact evidence is lacking in many claims and incoming dust seems to be the better explanation.)

Paleo-Indian stone artifacts like chert flakes from the end of the Pleistocene age have been found pitted with tens of thousands of particle tracks per square inch – and only on the side facing the sky. Similar evidence can be found on the moon. Analysis of moon rocks shows that micro-craters from dust particles and etched tracks from cosmic rays found in the glassy surface of many lunar rocks shows a peak in cosmic dust and solar flaring activity roughly 13,000 years ago. Thomas Gold analyzed the moon rocks for Cornell University and also concluded that glazed surfaces with droplets of melted and congealed rock suggest an episode of intense heating in which "the Sun's luminosity must have increased by as much as a hundredfold for a period of 10-100 seconds to have produced the observed effects" within the last 30,000 years.[201] This could be why Isaiah 30:26 warns "the light of the sun will be seven times brighter" and why so many ancient cultures claimed that cycles of destruction alternate between flood and fire. 2 Peter 3:6-7 says: "the world at that time was destroyed, being flooded with water. But by His word the present heavens and earth are being reserved for fire, kept for the Day of Judgment and destruction."

Even Mars has evidence supporting intense solar activity around the end of the last ice age. NASA's Mars Express team still finds new photographic evidence of catastrophic flooding on Mars every year. They often attribute the floods to very ancient times, but relatively young flood features often cut across older Martian craters. And near Mars' equator, at Elysium Planitia, a sea of frozen water has been found about 800 by 900 kilometers across, and about 45 meters deep. Liquid water cannot exist on Mars at current temperatures and pressures; even solid water ice sublimates (evaporates) away quickly and should not be accumulated in a frozen ocean – unless that ocean is so recent that it has not yet had time to sublimate away.

All of this can be explained by an incoming cosmic dust cloud during early historical times. Such dust would temporarily darken the earth while it energizes the sun, then heat and spectacularly brighten the sun. As it was described in ancient India: "Vishnu begins the terrible last work by pouring his infinite energy into the sun." I am reminded of Revelation 16:8 when "the fourth angel poured out his bowl upon the sun, and it was given to it to scorch men with fire." And in Isaiah 30:26 "the light of the sun will be seven times brighter."

The Utes tribe for which Utah is named has a myth in which the hare-god Ta-Wats shoots an arrow at the wayward sun god Ta-Vi, which leads to fire falling to earth and burning almost everything, including Ta-Wats, until his eyes burst from the heat and his tears flooded the world but extinguished the global fire. The Kutenai of Western Canada also have a myth in which a great fire consumes the earth after the sun is struck by an arrow. They also suggest a pole shift in the future, as they say "the end of the world is imminent" if the North Star moves from its proper place.[202]

Solar wind, which is usually strong enough to push out cosmic dust, would be temporarily overpowered by the stronger pressure-like forces of the galactic superwave, and the entire solar system would get incredibly dusty. Imagine watching a bonfire some distance away as a cloud of sawdust falls on the area. At first the dust would dim your view of the fire. Eventually the dust cloud would ignite or explode. Astrophysicists suggest that the sun would initially appear darkened even though it would be energized and put off more heat, until a solar blowout eventually removes all the dust from the inner solar system.

This may be exactly what happened around a star known as TYC 8241 2652. I first read about this star in an article from July 5, 2012 in <u>Science Daily</u> titled: "Dust Today, Gone Tomorrow." The article states: "An extraordinary amount of dust around a nearby star has mysteriously disappeared... This disappearance is remarkably fast... The dust disappearance at TYC 8241 2652 was so bizarre and so quick; initially I figured that our observations must simply be wrong." But incoming cosmic dust, energizing a star until it has a T-Tauri style coronal mass ejection, could be the explanation. T-Tauri stars are much like our own sun, except they emit more radiation and have more flaring. They put out much more infrared, about ten times more ultraviolet, up to twenty times more visible light, and about 100,000 times more X-rays, with intense and constant solar flaring that "generates a very strong stellar wind, expelling gas at speeds up to 1000 times that of the solar wind" we experience.[203] Five years later, scientists have no conventional explanation for the quick expulsion of dust, leaving us with Dr. LaViolette's theories, which have seen dozens of his expectations proven over the last 35 years...

The combination of forces of dust, radiation, gravity wave, and electromagnetic pulse reaches us from the galactic center, and the immediate aftereffects of it reaching our Sun (as solar repercussions could affect us as soon as a few minutes later) could be why be Egyptian mythology says: "Of itself the [Ra's] Eye is not strong enough to destroy them. Let it descend upon them as Hathor. So that goddess [Hathor, always drawn with the sun disk on her head] came and slew mankind." If our sun experiences a T-Tauri phase, Earth could temporarily get significantly hotter – enough to quickly melt polar ice whether it has relocated in a pole shift or not.

Dr. Paul LaViolette's <u>Earth Under Fire</u> discusses explosive phases of our galactic core, radiating outwards as galactic superwaves that trigger both extreme solar activity and pole shifts. Other authors agree with LaViolette that "the galactic superwave might even be a factor in making a star into a nova"[204] and that "the initial outburst from the galactic core caused electromagnetic shifts on earth, which may have caused crustal torque, pole shifts, tidal waves, and high winds. The first catastrophe was followed sometime later by an explosion of the sun's corona, caused by the influx of dust pushed by the galactic superwave."[205]

In Dr. LaViolette's own words: "superwaves periodically jerk the earth's axis…. The prompt arrival of the electromagnetic pulse and, some days later, the onslaught of the gravity wave, with its ensuing crustal torque, which would have caused earthquakes and volcanic eruptions…"[206] Despite his belief that the gravitational component of a galactic superwave event will only jerk the earth's crust a few meters – not thousands of kilometers (by itself, anyway, without considering the sun or the dust or the imbalanced ice caps) LaViolette also notes that "a farsighted society or group of individuals living thousands of years ago had valiantly attempted to communicate a warning message to Earth's future inhabitants about the past occurrence of a galactic superwave catastrophe."[207]

Since LaViolette cites ancient catastrophes, yet feels the gravitational component of his galactic superwaves are not by themselves sufficient to cause a pole shift of thousands of miles, I must assume a combination of forces at play. If the geomagnetic field is weakened and the viscosity of molten layers beneath the crust disappears along with friction, then even the small "jerk" of a few meters by the gravity wave could set the crust moving until mass imbalances are relocated or magnetic field strength increases.

The theories promoted by Schoch and LaViolette point out that the sun and the galactic center have been quiet and stable for thousands of years, but that it is normal for them to have periodic 'hiccups' when outbursts of gravity waves and intense radiation hit the earth hard enough to be the last straw that initiates a pole shift. (The last peak outburst of cosmic rays apparently reached Earth around 12,600 years ago, based on analysis of beryllium-10 concentrations in polar ice… and "analysis of this record indicates that these cosmic ray events recur about every 26,000+/-3000 years , approximating the duration of a precessional Great Cycle. On occasion, a 13,000-year half-cycle recurrence interval is also evident."[208]

As Hannes Alfven said in his acceptance speech for the Nobel Prize in Physics in 1970: "Space is filled with a network of currents that transfer energy and momentum over large or very large distances." Solar wind usually exerts enough outward pressure to keep most cosmic dust out, but when overpowered by an incoming galactic superwave, radioactive dust pushed into the solar system would have superconductive filaments which could transfer electrical charges between planets or between the sun and planets. There is much evidence of electric arc scarring on our moon, and on Mars, and such dust clouds and electric charge transfers could contribute to massive changes in solar output and flaring, and large and rapid changes in the Earth's magnetic field. It could also explain ancient references to gods who threw lightning bolts at each other. It could definitely help trigger a pole shift.

Analysis of several major supernova indicate they were triggered by the same outburst from the galactic center that would have reached us approximately 12,900 years ago when the Pleistocene extinction occurred and 'the ice age ended.' Perhaps it ended in some places like eastern Canada – which moved away from the North Pole that had been centered in Hudson Bay – but it started a new 'ice age' in places like northern Greenland and western Antarctica. Some believe this was the time of the last pole shift, and that they occur every twice in every cycle of precession (approximately every 12,900 years.)

One pole shift researcher with a focus on Antarctica is an Italian naval engineer, Dr. Flavio Barbiero. Long before becoming an admiral, he published Una Civilta sotto Ghiaccio (A Civilization Under the Ice) which may be the first book suggesting that Atlantis was in Antarctica – in 1974. "The earliest suggestion of Antarctica as the home of Atlantis seems to have come from a Chilean professor, Roberto Rengifo, who also proposed, in 1920, that Antarctica was the original home of modern man until a catastrophic pole shift forced migration northwards."[209] But Rengifo did not write an entire book about Atlantis in Antarctica.

Barbiero did write a whole book on the subject, and he clearly said that a very rapid pole shift moved Antarctica from a temperate latitude close to the South Pole. "The only way to completely and coherently explain what took place at the end of Pleistocene appears to be that of admitting the possibility of a shift of the poles of the same magnitude Hapgood hypothesizes, but in a much shorter time: not more than a few days." Barbiero suggests that only an extraterrestrial input of energy, like a galactic superwave or asteroid impact, is capable of applying enough torque to the crust of the earth to cause "almost instantaneous changes of the axis of rotation and therefore instantaneous shifts of the poles in any direction and of any amplitude." (He estimates, however, that a twenty degree shift seems most likely – with maximum changes in elevation of approximately half a mile.)

Barbiero argued that survivors from Atlantis founded new civilizations on coastlines around the world, but that the first few thousand years of evidence would now be submerged under at least 100-200 feet of ocean. For the first thousand years after the last pole shift, coastal settlements were destroyed by rising sea levels as the former ice caps melted faster than ice accumulated at the new poles. Sea levels rose at least 300-400 feet (largely in one event called "meltwater spike 1A" right around 12,900 years ago) before levelling off. By the era of the evidence of civilization we do find, the identical (Atlantean) nature of the far flung groups has evolved and diverged and the shared Atlantean heritage of known civilizations is not as obvious.

Another pole shift author with a focus on Atlantis in Antarctica is Rand Flem-Ath. In 1976 he noticed that ancient maps like the one made by Athanasius Kircher in 1665 seem to show Antarctica's sub-glacial contours. In the course of his research he soon found the Piri Re'is map of 1513 and the books of Charles Hapgood. In 1977 they began corresponding, and Hapgood wrote: "I am astonished and delighted by your article which arrived here today. Believe it or not, it is the first truly scientific exploration of my work that has ever been done. You have found evidence for crust displacement that I did not find."

Flem-Ath and Hapgood continued writing each other up to Hapgood's fatal "accident" in 1982. Flem-Ath went on to author or co-author several books relevant to pole shifts, including: <u>When The Sky Fell: In Search of Atlantis</u>, (which he wrote with his wife Rose) <u>The Atlantis Blueprint</u>, (which he wrote with Colin Wilson) and <u>Atlantis Beneath the Ice</u> – all promoting his conclusion that Western Antarctica was Atlantis, destroyed and moved into a polar region during the last pole shift. (Eastern or Greater Antarctica is polar now and was also within the Antarctic Circle during the previous orientation of the surface of the planet – when the North Pole was in Hudson Bay, the South Pole was just off the Indian Ocean side of Antarctica and Western or Lesser Antarctica, near South America, was ice free with a temperature climate similar to that of modern Britain.)

"Ancient maps represent the earth as it was before the last earth crust displacement, when North America was under ice and a third of Antarctica was ice-free."[210] Flem-Ath agrees with Hapgood's theories on ancient maps and that the last North Pole was in Hudson Bay, though Flem-Ath believes in a very long 41,000 year cycle and assumes the pole was there from 50,600 B.C. until 9,600 B.C. (On the one hand, this date would coincide with Plato's date for the destruction of Atlantis, which is supportive. But it also coincides with the sudden rise in temperature in seawater in the North Atlantic, leaving no room for intervening years when the old ice cap melted.)

Flem-Ath also points out that Antarctica shares many of the features of Plato's Atlantis: mountainous with a high average elevation, in the middle of the world ocean, larger than Libya and Asia (Turkey) combined, with little islands between it and another continent. If Charles Hapgood had not been killed in 1982, it is possible that Flem-Ath would have co-authored a book with him about the "convincing evidence of a whole cycle of civilization in America and Antarctica, suggesting advanced levels of science" that Hapgood wrote to him about shortly before he died.

The pyramids at Giza in Egypt are also starting to show evidence of very ancient civilization. In my opinion, we have been taught lies by mainstream Egyptologists like the former Egyptian Minister of Antiquities, Zawi Hawass. He denies every relevant fact that would prove the pyramids were never built as tombs for pharaohs, and that they are over 10,000 years old. He has said "Carbon-dating is useless. This science will never develop. In archeology we consider carbon-dating results imaginary."[211] On DNA testing Hawass said: "It is not always accurate and it cannot always be done with success when dealing with mummies. Until we know for sure that it is accurate, we will not use it in our research."[212] At a conference in Cairo on April 22, 2015, Hawass disputed discoveries made underground using radar, and said: "I don't believe in radar." Hawass simply dismisses all scientific evidence that questions mainstream

Egyptology's conclusions or supports very ancient dates for Egyptian monuments.

Nicotine and cocaine (native only to South America) found in almost every mummy cannot be denied. The star map carved into an eight inch tunnel only accessed by a robot camera (in an eight inch "air shaft" in the Great Pyramid near "Gantenbrink's Door") for the first time in the 1990s shows the sky as it looked over 11,000 years ago. It had to be carved in place before construction – no one can fit in the shaft after that… And despite the meticulous recordkeeping of the ancient Egyptians, Hawass' American partner in crime, Mark Lehner (with whom he just wrote Giza and the Pyramids: The Definitive History) once said: "Although we are certain that the Sphinx dates to the Fourth Dynasty, we are confronted by a complete absence of Old Kingdom Texts which mention it."[213] It seems kind of funny that this "definitive" book has no mention of the giant caverns inside the Great Pyramid, like the void discovered near the Kings Chamber that received much publicity in November 2017. The "ScanPyramids Big Void" seems to be about 98 feet long and 50 feet high. What secrets lie within? We won't know anytime soon – authorities like Hawass have kept almost all investigators away from the Great Pyramid since 1995 and from the Sphinx since 1993.

Very interesting to me – this Egyptian and American duo, Hawass and Lehner, are the most respected Egyptologists despite their educations both being funded by the A.R.E. – the Association for Research and Enlightenment founded by the American psychic Edgar Cayce, who suggested that we could find a Hall of Records from Atlantis underneath the Sphinx. The main A.R.E. goal in Egypt, and one of its main goals overall, is to find the Hall of Records from before the last pole shift. (The A.R.E. also published the last book I found with a true focus on pole shifts prior to my own book you are reading now – they published John White's Pole Shift: Predictions and Prophecies About the Ultimate Disaster in 1980.) Despite the source of funding for his education, Hawass denies the existence of tunnels and chambers under the Sphinx, even though sonar has detected a chamber beneath its paws just where Cayce said an entrance could be found – and there are many photos available online showing Hawass entering descending shafts by the head and the rear of the Sphinx. As Barbara Hand Clow wrote in Catastrophobia back in 2001: "There are persistent and well-founded rumors that the Egyptians are carrying on clandestine digging all over the [Giza] Plateau and that many things have already been found but kept secret."[214]

I think it is very possible that the measurements built into the Great Pyramid contain scientific and mathematical knowledge related not just to the earth, but also to earth's distance from and cycles involving the moon, sun, and galactic center. Graham Hancock, author of many great books including: <u>The Message of the Sphinx</u> and <u>Fingerprints of the Gods</u>, lists these and many other measurements hidden in plain sight, built into the pyramid. Scott Creighton suggests that we consider the angles of the "air shafts" or star shafts in the Great Pyramid. There are two northern ones and two southern ones, and they just might represent the angles of the highest zeniths of Orion's belt stars many thousands of years ago: "The star-shafts of the highest chamber, however, seem to indicate a time when the heavens were out of balance and had rapidly changed... These two figures (45.1º and 71º) present to us the pre and post pole shift angular distances of the star Al Nilam at the Giza local horizon. The difference between these two values will then tell us the distance this star shifted in the heavens after the pole shift event."[215]

Peter Lemesurier, in <u>The Great Pyramid Decoded</u>, also noted that the distance from one corner of the pyramid to the opposite (the diagonal of its base) – the distance one needs to cross to end up half way around the pyramid at the opposite corner, is approximately 12,828 inches. "Of all the alignments associated with the Great Pyramid, the 'pyramid diagonals' are among the most precise."[216] Perhaps there are reasons for the extra care the builders took to keep that measurement accurate. The length in inches is remarkably close to estimates for one half precession cycle in years, and a great many correspondences of "pyramid inches" to "prophetic years" between events can be found throughout the pyramid.

The geologist Robert Schoch was able to investigate before most restrictions were put into effect – but his findings led to the restrictions. Mainstream Egyptologists don't want a very ancient alternative history to turn their version of events upside down. Major religions don't want that either. Evidence that disproves anything in the Qu'ran seems unlikely to be approved in a Muslim nation like Egypt. The Vatican also doesn't want knowledge of cycles of catastrophes to replace the once-and-done idea of redemption through Jesus Christ. And assuming these cycles of destruction are predictable, major governments don't want the general population to know when the next event is either...

Despite such forces working to push an incorrect and far too recent construction date for Egyptian monuments, Dr. Schoch has helped to disprove this alleged 2,500 B.C. time frame. Dr. Schoch is most famous for proving the Sphinx in Egypt is many thousands of years older than we were taught, (because he proved the erosion on the Sphinx is from massive rainfall, not blowing sand – and must have existed when the climate was still wet and rainy) making it at least 10-15,000 years old. At first, such an old date for the Sphinx was countered with the comment that there were no signs of the civilization that could have built it at that early point. But now the site of Gobekli Tepe in Turkey has been conclusively dated to be at least 11-12,000 years old by the German Archeological Institute.

As for what may have caused the demise of civilization so far back – Dr. Schoch generally agrees with Dr. LaViolette, and said in his recent book (Forgotten Civilization) that: "the last galactic superwave hit Earth around 12,900 years ago." Others think comet impacts are responsible for every major disruption of the earth's surface. Some researchers think that a comet hit North America around 12,000 years ago. Even if a comet or its fragments did hit the earth at that time, I do not think it was the cause of the pole shift that ended the ice age in North America. There are too many ancient references to predictable cycles and usual intervals and there is much evidence for catastrophes arriving after specific intervals. Comet impacts, however devastating and however often they may be, are unpredictable, one-time events.

I agree there is compelling evidence, like Burckle Crater in the Indian Ocean, suggesting that Noah's Flood was caused by a mega-tsunami from a comet impact about 5,000 years ago. I even take notice of some noteworthy parts of a "myth" from some of California's native tribes. The Cahto tribe says that at the Great Flood: "The sky fell. The land was not. For a great distance there was no land. The waters came together. Animals of all kinds drowned."[217]

The Yurok tribe says: "The sky fell and hit the water, causing breakers that flooded all the land... Sky-Owner sent a rainbow to tell them the water would never cover the world again."[218]

This sounds a lot like the "rainbow covenant" in the biblical story of Noah, assuaging fears that the Great Flood was a one-time event (from a comet impact) and not from anything that will happen repeatedly in predictable cycles, like a pole shift caused by periodic waves of energy from the galactic center. I know the popular Christian view of the "rainbow covenant" in Genesis is that there will never be another flood to almost destroy mankind... But many biblical references to cyclic destruction by flood – along with a great deal of non-biblical evidence of pole shifts – make me quite certain that we can expect these periodic catastrophes again. Consider the Catholic Bible's 2 Esdras 3:9-18 "But again, in its time you brought the flood upon the inhabitants of the world and destroyed them.... You bent down the heavens and shook the earth, and moved the world, and caused its depths to tremble.... Your glory passed through the four gates of fire and earthquake and wind and ice." This sounds like a description of a pole shift and flood (happening AGAIN) to me. I suspect this is the correct analysis; and that the last pole shift was about 12,960 years ago, and the previous one was about 12,960 years before that...

One good video presentation about these recurring catastrophes can be found under titles like "The Galactic Cross is Upon Us" or "Cosmic Patterns and Cycles of Catastrophe," many of which follow presentations by Randall Carlson on ancient sacred geometry. These videos repeatedly show an image like a circle and a cross, in which the circle is the 25,920 year cycle of precession, divided into twelve sections or "Ages" of 2160 years for each of the zodiac constellations. As history advances over the last 150,000 years and spirals around this circle of repeating ages, the video notes that almost all the great extinction level events occur on one straight line, one axis through this cycle – either at the start of the Age of Leo (12,960 years before or after our present time) where five of the top ten events occur – or at the start of the Age of Aquarius, where three of the last top ten events occur, including the Greenland Blitz, (104,000 years ago) the Heinrich Events of 52,000 years ago (when Neanderthal populations were decimated) and the onset of the Late Wisconsin Ice Age 26,000 years ago, (when Neanderthals went extinct.)

We are entering the Age of Aquarius again now (a vague start date, but some astronomical societies use 2010 as its approximate start date.) Perhaps we should not ignore the fact that the image of Aquarius shows the gods pouring out a flood, as we may be due for another Great Flood very soon. It would seem that Earth's cycle of precession of the axis through the zodiac is linked to a cosmic cause which is also linked to or causing repeated cycles of catastrophe. As Dr. Paul LaViolette wrote: "Perhaps the Earth's precessional period has become entrained by the heartbeat of the Galactic core."[219]

I have listened to many impressive discussions with Randall Carlson in dozens of videos and interviews… in one lecture he says: "There seems to be a tempo in the delivery of cosmic matter to the inner solar system. It doesn't seem to be random… The delivery of cosmic material and energy – the energy pulses that would be affecting Earth are non-random. They're on some kind of cosmic timetable, a cosmic tempo if you will. I think this is one of the most important insights we get from these ancient traditions – the measurement of cosmic time and how it relates to us here on Earth."

In another video from 4/24/17 he says: "The paradigm has to shift. We have to realize we are inhabiting a planet that is a hell of a lot more dynamic than anybody has imagined… these catastrophic events are part of the norm." In yet another video he says that "we, even twenty or thirty years ago, did not appreciate the magnitude of these catastrophes that have constantly re-occurred throughout the last couple hundred thousand years that we humans have been occupying this planet."[220] He says climate change is natural and we shouldn't be worried about the small amount of global warming some want to attribute to greenhouse gases, when there is evidence of previous temperature changes up to about 10 degrees Centigrade (18 degrees F) in as little as a year or two caused by exogenic (cosmic) events…

Carlson's impressive conclusions and diagrams remind me of Plato's comments on "a pattern set in the heavens, where those who want to see it can do so"[221] and several other clues to the timing of pole shifts… In the stories of ancient India, the god Vishnu returns in many forms in many ages, but he always avenges evil and rebuilds destroyed worlds as if on some predetermined schedule. As the authors suggest in Hamlet's Mill: An Essay Investigating the Origins of Human Knowledge and its Transmission Through Myth "It is this regular returning of avatars of Vishnu which helps clarify matters. Because it is Vishnu's function to return as avenger at fixed intervals of time..."[222] Keeping in mind that "the galactic center is called 'Brahma,' the creative force, or 'Vishunabhi,' the navel of Vishnu"[223] we can safely assume there is a reappearance of the visibly active galactic center at fixed intervals of time linked to world-age-ending disasters.

The Rig Veda also describes this as the "seat of Rita" where Svarnara is "the celestial spring… which Soma selected as his dwelling."[224] Soma is "lord of the poles." So in this alternate description from ancient India, the galactic center seat is a throne for the lord controlling the poles of the earth and a heavenly spring or source from which heavenly waters of matter and energy flow out.

The seat of Vishnu is also often depicted as Amritamanthana, the "Churning of the Milky Ocean" or the "mighty churn of the Sea of Milk" and much ancient Indian artwork shows Vishnu seated above what appears to be a butter churn or a fire-drilling stick. Two sets of opposing forces of good and evil pull back and forth on a serpent-rope coiled around the axis of the churn. This suggests a seat at the center of the Milky Way and a regularly repeating change in the direction of rotation.

Very similar drawings exist in Egypt, depicting Horus and Set, the son and brother of the deceased highest god Osiris, drilling or churning a giant axis by pulling on opposite ends of a rope coiled around it. Set murdered Osiris and represents chaos and destruction and evil, and Horus represents the good son's fight to restore order. At Angkor Wat in Cambodia, 540 statues all pull on a serpent over a water ditch (the Milky Ocean?) with a type of Mount Mandara churning staff in the middle. In Mexico, the god Tezcatlipoca used a fire-drill to kindle new stars just after the Flood. Another similar drawing in the Mayan Codex Tro-Cortesianus shows opposing deities pulling on a serpent rope that goes through an hourglass shaped churn and a turtle – representing the earth – with a glyph of the sun on the rope at the churn, possibly indicating that such churning events are noticed when the winter solstice sun appears near the galactic center.

In Finland, fire is started in the "cradle on the navel of the sky" – but is this near the galactic center, or Polaris – the pole star? Of course, after a pole shift initiated by an event at the galactic center, and the creation of a new world with a new axis and a new pole star – the two locations are both integral to any story of the destruction and re-creation of the world...

The Bhagavita Purana describes "...the exalted seat of Vishnu, round which the starry spheres forever wander, like the upright axle of the corn mill." Some readers might assume this is as likely to be a description of the pole star – which stars appear to rotate around – as the galactic center – which modern science has realized the stars actually do rotate around. But the distinction is clarified in the Hindu story of Dhruva, in which Dhruva's mere contemplation of the galactic center as the seat of Vishnu made the Earth bend and sink down, after which Vishnu transformed Dhruva into the pole star, Polaris – but only for a fixed span of time. Verse 2:8 of the Vishnu Parana clarifies further that Dhruva is at the summit of Meru, the World Mountain in the North, as is Polaris, the nail in the toe of Vishnu's foot. (The North Pole is often referred to as the north nail in myths around the world.) Vishnu's seat, however, is a different location in a river, presumably the Milky Way, as other writings indicate Vishnu's seat is the galactic center.

There is a transference of energy from the galactic center to the North Pole in the Bhagavita Purana: "The river flowed over the great toe of Vishnu's left foot, which had previously, as he lifted it up, made a fissure in the shell of the mundane egg, and thus gave entrance to the heavenly stream." The mundane egg is definitely the Earth. Action from the galactic center (lifting Vishnu's foot) took away a stabilizing power source from the North Pole and the spinning Earth. But what then flowed from the galactic center into the North Pole and cracked the world?

In the theory I describe, the galactic center becomes active, periodically spewing out dust, light, radiation, gravity waves and other forms of energy in what Dr. Paul LaViolette named a "galactic superwave." This becomes visible upon arrival in our solar system about 23,000 years later, illuminating the galactic center as the superwave starts to push dust into our solar system, darkening and then energizing the sun, while a gravity wave jerks the earth and causes a pole shift. An electromagnetic current of ionized particles would flow in glorious huge auroras down into the magnetic poles of the earth.

The authors of Hamlet's Mill ask of ancient myths: "Why does it always happen that this Mill, the peg of which is Polaris, had to be wrecked or unhinged? Once the archaic mind grasped the forever-enduring rotation, what caused it to think that the axle jumps out of the hole? What memory of catastrophic events has created this story of destruction? …Why had Dhruva to be appointed to play Pole Star – and [only] for a given cycle?"[225]

Unfortunately, the Hindu commentaries on this are difficult to understand for the majority of us who have not been trained as initiated priests in their traditions. Much like followers of Hermetic Wisdom are taught not to throw pearls before swine, and how Jesus often spoke in parables and made it clear that only those with eyes to see and ears to hear would understand completely, the Rig Veda also acknowledges levels of meaning hidden in "secret words" (Verse 4.3.16) and that there are multiple levels of understanding revealed to initiates:

"Four are the grades of speech that have been measured;
Men of the divine knowledge who are wise know them.
Three of these kept in secret make no motion,
[Most] People speak only the fourth grade of speech." (Verse 1.164.45)

Modern Western thought may focus on linear progress, but almost every ancient tradition understood history as a series of recurring events in regular, predictable cycles, with the Great Year of the Gods (the 25,920 year cycle of precession – and catastrophes) being the longest one to worry about…

Carlson's videos seem to back up Dr. LaViolette's analysis that "the ice core beryllium-10 record indicates that galactic superwaves pass us about once every 26,000 +/- 3,000 years, approximating the period of one polar precession cycle, with the possibility of a 13,000 year recurrence interval [another superwave event, possibly less intense, halfway through each cycle.] This periodicity [which we clearly detect in the 25,920 year cycle of precession of the Earth's axis] may originate in the mechanism that generates superwaves at the Galactic core."[226]

Carlson's videos also reminded me of (but never mentioned) a book called The Mysteries of the Great Cross of Hendaye: Alchemy and the End of Time – by Jay Weidner and Vincent Bridges. Their book describes an old stone monument in Hendaye, on the border of France and Spain – and a chapter in an occult book (Mystery of the Cathedrals) written under the pseudonym "Fulcanelli" (The Sons of Vulcan) in 1926. Fulcanelli explained that cosmic disasters happen in a predictable cycle, and that alchemy was never about creating gold from lead, but was a veiled allegory about properly restoring a golden age after the next cataclysm. Fulcanelli said that secret societies have been passing down knowledge of these cycles for thousands of years, and that the monument at Hendaye is just one of many monuments in stone that "pointed to a specific time period, the intersection point of several celestial cycles" regarding the galactic center.

"The symbols and teachings encoded in the cross at Hendaye offer us a new understanding of the cosmos, especially with regard to the center of the Milky Way Galaxy and its effect on us... ancient knowledge of the location of the center of our galaxy and from that knowledge a way to estimate the date of a celestial event of eschatological magnitude."[227] He also warns that a "double catastrophe" for the world is due soon, "when the solstices cross the galactic axis – the time, according to Fulcanelli, of the cyclic catastrophe"[228] which Weidner and Bridges interpret as Judgment Day, due in the early 21st century.

"The starlike disturbance from the center of our galaxy might trigger a reaction in our own sun, hence Fulcanelli's insistence on a 'double' catastrophe."[229] (This could be why the Greeks feared both destruction by fire (ekpyrauses) and destruction by ice (kataklysmos) or water.) Weidner and Bridges also discuss "politics of secrecy surrounding the knowledge of this oncoming celestial event" and that factions of the secret societies who maintain this ancient wisdom have been arguing for a thousand years over how to handle the knowledge.[230] "Graham Hancock, in his groundbreaking work Fingerprints of the Gods, presented a catalog of cataclysmic events, and concluded that some sort of upheaval and flood occurred around 13,000 years ago... the cataclysm –

which in his view resulted in a massive shift in the earth's crust – he does suggest that it is somehow related to the precessional cycle and its cosmic clock."[231] Hancock wrote that "'the essence of the cult (the secret of the catastrophe's timing) might survive, carried forward by a nucleus of determined men and women.' Their objective would have been to preserve this knowledge for a future civilization facing the same event."[232]

Macrobius warned that "those who have learned the mysteries should hide the unsearchable secrets." Plato, Herodotus, and countless other ancient wise men said pretty much the same thing – that the initiated are under oath to never reveal such mysteries clearly. But Weidner and Bridges suggest that "not only have the initiated few survived for centuries, right down into our own time, but apparently some of them wanted to reveal the secret as well... one of the initiated gave away the secret in 1957..."[233] in the form of "Fulcanelli" adding a chapter to Mystery of the Cathedrals.

The evidence for such pole shift disasters in our past comes from many different fields of science. Coral reefs circle the earth in a band around the equator – but older 'equators' of coral reefs can be found crisscrossing the earth everywhere – even through what is now the arctic. Mammoths with undigested summer vegetation in their bellies were suddenly frozen. There is evidence of 'ice ages' in India and Africa – yet tropical species survived, so the entire planet could not have iced over. Magnetized ions in lava rock indicate where magnetic north was located when the lava solidified – and analysis sometimes proves that huge changes occurred in a short time. India warns us that Vishnu the Destroyer returns at regular intervals. Egyptian priests told the Greeks these events occur "after the usual interval." Randall Carlson and Paul LaViolette suggest the dreaded events occur every half-precessional cycle. Many sources suggest the next pole shift is due "soon" – in the early 21st century. "Fulcanelli" may have given a warning showing that initiated brotherhoods preserving ancient knowledge may have members who are willing to spill the beans and reveal the timing of the event, even if done in a manner that would only leak out slowly through publications few people would read or understand.

What about evidence in publications that almost everyone is familiar with? Is there evidence we have already seen and overlooked right in front of our faces – in the Bible?

Pole Shifts in the Bible

Professor Hapgood worked for the OSS (the precursor to the CIA) during WWII. He should have known that he was risking his life if he expressed a willingness to spill the beans and publicize exciting new evidence on pole shifts and previous civilizations. The ideas Chan Thomas wrote about and the CIA classification of his work also remind me of various pole shift references that the Roman Catholic Church apparently wanted to keep "classified" by keeping most of the books that mention pole shift evidence out of the official version of the Bible. At poleshiftnews.com they write: "98% of the human habitants on Earth are totally unaware of what is taking place right now. Is that by design?"[234]

The Council of Nicaea decided that certain books would not be included in the official canon in the year 325, and anything that did not help support the authority of the Church or the most important doctrines of the new religion were cast aside. A main premise of Christian thought is "once and done" linear progress from sin to salvation through the single Messiah Jesus Christ, not an endless cycle of repeated worlds in which "the earth" is destroyed over and over such as Hinduism describes with returning avatars of Vishnu at regular intervals.

To support the central ideas of Christianity, sometimes unwanted books had to fall to the wayside. The Book of Jasher is mentioned in the Book of Joshua. Jasher 88:64 describes Joshua's Long Day, in which the sun stood still and lengthened a day of battle so that Joshua could more completely annihilate the enemies of the Jewish people. "And the Lord harkened to the voice of Joshua, and the sun stood still in the midst of heaven, and it stood still six and thirty moments, and the moon also stood still and hastened not to go down a whole day." I can only think of a few explanations for such an event: it is fictional and never happened; it was God's miraculous divine intervention; or as many people have suggested: it describes the strange interruption of the normal patterns of heavenly bodies during a pole shift.

There is another biblical figure that the church does not want in the Bible, quite possibly for the same reasons. In 2014 I wrote a few articles detailing why the Book of Enoch might not have been included in the Bible – because it has too many references to pole shift evidence.

"Enoch and Elijah – Witnesses to POLE SHIFTS"[235]

"The early theologians Irenaeus and Hippolytus may have been the last heirs to the uncorrupted oral tradition of the Apostles, and both of them said that Enoch and Elijah were the two witnesses we will see in the future preaching and dying on the streets of Jerusalem in the end times. The Bible tells us that their presence will be a sign that the Second Coming is imminent. I believe their stories also tell of pole shifts in our past, and lend additional support to the idea that they will be a witness to an upcoming pole shift, when we receive "a new heaven and a new earth."

The Bible tells us that Enoch lived a total of 365 years. This lifespan brings two obvious ideas to mind: first, that no one normally lives that long – and second, that the years of Enoch's life match the number of days in a year. If this is meant to draw our attention to the sun and the number of days it takes for the earth to orbit the sun, what else might we take from the story of Enoch to apply to the sun?

Enoch's life is broken down into two portions: at the age of 65 he fathered Methuselah, then continued to live another 300 years before he "walked with God" and was "translated away so that he did not see death." (Hebrews 11:5) The word "translated" comes from the Greek "metatithemi" which means "taken to another place." Because I have spent many years researching topics like pole shifts, this makes me think the story of Enoch may incorporate a coded message about a pole shift event in the distant past when the sun (symbolized by Enoch in this story) would have been positioned in the sky next to the galactic center (God) when it suddenly shifted its apparent position in the sky and was "taken to another place."

I also suspect that the way the Bible breaks down Enoch's life into two unequal portions might be similar to the way the Maya broke their year down based on the dates on which the sun passes directly overhead in the zenith. Every spring equinox in March, the sun is directly above the equator at noon. Lands north of the equator and south of the Tropic of Cancer witness the sun directly overhead at noon at some point prior to the first day of summer, and again a second time of equal duration after the solstice. For example, at Chichen Itza, where the Maya built the great Pyramid of Kukulkan, the sun is directly overhead on May 22. The sun appears to keep moving north until summer, then starts to fall southwards again. Chichen Itza experiences a second solar zenith on July 19. At another Mayan city (Izapa – which is farther south, at a different latitude) where this pattern was first recognized, the two zenith dates are 105 days apart, and the Maya there broke the year into 105 and 260 portions based on the spread of the zenith passage dates at Izapa's latitude. What latitude, I wondered, would experience zenith passage dates 65 days apart – and could such zenith dates possibly point us to the previous position of Jerusalem, before the last pole shift?

A little research led me to a chart labelled "Zenith passage dates of the Sun for Observers in Different Latitudes" in Anthony Aveni's Skywatchers of Ancient Mexico. Just below 20 degrees north there are 65 days between zenith dates. Could this be the latitude the holy site we now know as Jerusalem used to be located at prior to the last major shift of the earth's crust relative to its axis of rotation? Could the location of this ancient holy site possibly even explain why the Maya had a special term for the 65 day period (Aveni calls it the Cociyo) when neither Izapa nor Chichen Itza experience zeniths 65 days apart?

At first glance, this seems to be a dead end and a bad theory, because we know from other data that the previous North Pole was located on the west side of Hudson Bay. When the North Pole was there, the land that is now called Jerusalem was approximately eleven and a half degrees north latitude, not just under twenty degrees north latitude. Jerusalem, at its former latitude, had zenith passage dates well over 100 days apart. However, I soon noticed that the change in latitude – the northward movement Jerusalem experienced as a result of the last pole shift – is approximately 19.8 degrees. Which means that Enoch's being taken away in a "translation" and the years of his life may be astronomical references to a pole shift after all.

I believe that cosmic events emanating from the galactic center cause recurring, periodic, and predictable pole shifts on earth. At the same time I believe the sun will appear to be dark (Revelation 6:12 "the sun became black as sackcloth") for three days, at the point in the year when it appears at the crossing point of the galactic axis and the ecliptic – the apparent path of the sun and planets. I suspect it is no coincidence that Jesus died on a cross and was dead for three days and that we are told the sun went dark when he died. Jonah was in darkness in the belly of the whale for three days, and I believe this is a reference to the sun going dark as it passes through the cosmic leviathan of the Milky Way's central bulge along the galactic axis. Even Native American Indians have similar tales. The Abenaki tribe in New England tells a story in which Gluscabi warns the animals: "A terrible thing is going to happen. The sun is going to go out. The world is going to end and everything is going to be destroyed."[236] The Nisqually tribe of the Pacific Northwest have a legend of massive earth changes and mountain formation followed by mankind's loss of fire (which is an almost universal symbol for knowledge/technology) after which "the people saw the sun rise for the first time in many days."[237]

If I am correct, for the first time in many thousands of years, the sun goes dark at approximately the same time that the galactic center becomes visibly illuminated. I am reminded of Christ's death and resurrection with three days in the tomb of the earth, and with the Egyptian story of the sun-god Horus resurrecting his father Osiris: "it was this sacrificed left Eye of Horus, when presented as an offering to the mummy of Osiris, that restored the deity to life."[238]

Even Islam expects three days of darkness at the end of the world when the Tariq Star appears. In Islam's "Final Signs of Qiyaaamah" we are warned: "The ground will cave in... Smoke will cover the sky... A night three nights long will follow the days of smoke-clouded skies. After the night which is three nights long, the next day the sun will rise in the West." Muslims are also told that Allah "revives the earth after its death" in another reference in the Qu'ran (Surah 30 verse 19.) This certainly sounds like a pole shift to me...

Getting back to the Bible, Elijah encountered "a chariot of fire." The Book of Enoch mentions "the sun... and the chariot in which it rises." (Enoch 72:5) We can safely assume Elijah represents another heavenly body which encountered the sun. "And Elijah went up by a whirlwind into heaven" just after he walked past Bethel (Beth-El = the house of God = the galactic center) and crossed the

Jordan River (the Milky Way) and encountered the chariot of fire (the sun.) So this occurred when the sun (and some other heavenly body – I suggest Jupiter) had just passed the galactic center and the axis of the Milky Way. 2 Kings 2:17 describes the men sent out to look for Elijah after he disappeared – "They searched three days but could not find him." Another symbolic reference to the three days of darkness and confusion.

The Egyptians described their sun-god Ra, riding in the Boat of Millions of Years, when he "entered at the head of his holy mariners and established himself on the throne of the two horizons. The holy one had grown old, he dribbled at the mouth, his spittle fell upon the earth… and the sacred serpent bit him. The flame of life departed from him." This describes the sun at the galactic center, by the stinger of Scorpio, with the sun going dark as a cosmic (spittle) dust cloud is pushed into our solar system. The goddess Isis tries to heal Ra, but "the divine one hid himself from the gods, and the throne in the Boat of Millions of Years was empty."

Of course the sun doesn't stay dark forever. In Mayan tradition we simply enter a new sun, a new world age. John Major Jenkins wrote about Mayan mythology and how it is expressed in their monuments and their ballcourt game. In their myths, the "Hero Twins" resurrect their dead father, the winter solstice sun One Hunapu. Resurrecting the dead father sun has roots around the world, as with Horus and Osiris in Egypt. But the Mayan Hero Twins may represent the North Pole and South Pole, as portrayed in Hopi Indian legends… First the Hero Twins must go on a journey (if the poles move – a pole shift) to shoot the false god Seven Macaw out of his sky perch. Seven Macaw is the polar constellation we know as the Big Dipper. He appears, falsely, to rule the sky from a central North Pole throne around which the stars seem to rotate. This sounds much like the biblical Satan – as Isaiah 14:13 warns he said: "I will ascend to heaven; I will raise my throne above the stars of God, and I will sit on the mount of assembly in the recesses of the north."

It also shares some common ground with India's Bhagavata Purana and the tale of Dhruva being made into a pole star, where it will seem to the unwise like the skies forever revolve around him. But he is only allowed to be the pole star until the end of the age, after which he (being a good and divinely appointed polar ruler, unlike Seven Macaw) ascends to what appears to be the galactic center, the highest heaven, "to the exalted seat of Vishnu, round which the starry spheres forever wander, like the upright axle of the corn mill circled without end." The Mayan true center is the "Heart of Heaven" (again, presumably the galactic center) and eventually Seven Macaw is known to be a false god, his North Pole throne is known to NOT be the highest throne of heaven, and like the biblical Satan, is cast down before the revival of the Maya's true high god, One Hunapu, the winter solstice sun in conjunction with the galactic center.

This alignment is represented by the Maya ballcourt game in which victory comes when the ball is finally put through the hoop – "because the characters of Maya myth have astronomical identities, the ballgame ultimately describes specific celestial processes and events."[239] It represents the cycle of precession and the winter solstice sun finally going through the galactic center after thousands of years of waiting for what happens now (with the winter solstice sun's disk completely passing the galactic center for the first time on December 21 approximately in the year 2019.) Jenkins summarizes the Mayan message as: "The gameball is in the goal, the game is over, Seven Macaw is dead, the solstice sun can be reborn."[240] As Dr. Paul LaViolette noted: "This solstice position was so important to the Maya… the Maya were aware that this Galactic center region has a major influence on terrestrial cataclysmic cycles."[241]

In Norse mythology, at the final battle of Ragnarok, the sun dies but gives birth to a daughter. The Egyptians simply replace Ra with Osiris. Osiris became king of the gods "on that day of the union of the two earths… in the double horizon." In the Egyptian Book of the Dead, we also learn that Osiris dies at a similar occasion: "on that day of the union of the two earths, the gathering of the two earths it is at the sarcophagus of Osiris." Osiris was replaced by his son Horus, specifically "Horus of the two horizons" – who is eventually stung in the foot by a scorpion and Horus is reborn as Osiris. The cycles continue…

Similar themes come up in the Christian Bible. Revelation 3:21 may describe the same death and renewal of world ages with the sun at the galactic center when it says: "I will grant to him to sit down with Me on My throne, as I also overcame and sat down with My Father on His throne." Egyptian ideas are also reflected in the reference in Genesis 3:15 when God informs the serpent he will only be able to bruise the heel of the future messiah, but not permanently overcome Him with death – again indicating death at the same spot near the galactic center/the stinger of Scorpio. Christ was only temporarily dead for three days, and he also represents the sun.

Going back to the three days of darkness idea... One of the even stranger comments in the Bible appears in 2 Kings 2:24 "Two female bears came out of the woods and tore up forty-two lads" from Bethel who had previously mocked Elijah's son. Now I have been to that part of Israel. I have stood in the Jordan River. Trees are scarce, and I never saw "woods" or a forest dense enough that bears could emerge and suddenly surprise anyone. And even if we grant that there had been thick woods and a pair of real bears, can you imagine 42 young men not scattering and running off in different directions? Surely a bear could kill a lad or two, but forty-two of them?

As C.M. Houck commented in The Celestial Spheres: Keys to the Suppressed wisdom of the Ancients: "How could two bears possibly manage to outrun, catch, and destroy forty-two terrified, hyperactive juvenile delinquents? They couldn't. This is sacred language."[242] And what I think he means is that this is another astronomical reference, this time to the two "polar bears" – the constellations Ursa Major and Ursa Minor – the Big Bear and Little Bear near the celestial North Pole. I suggest that during the last pole shift, it was noticed that these "bears" seemed to suddenly move far faster than usual into the sky, corresponding with the sudden disappearance of 42 visible stars (or forty two degrees) which unexpectedly fell below the opposite horizon...."

UPDATE: In December 2014 I came up with more evidence that Enoch CLEARLY wrote from a location at approximately 50 degrees latitude – nowhere near the Middle East, and quite possibly from the capital city of Atlantis.

My next closely related article was titled: "Enoch Lived in Atlantis"[243]

"In the Book of Enoch – Noah's Grandfather – Enoch wrote about a variety of subjects, including the heavens. In one part, he describes the annual changes in the length of days and the relative balance between night and day. Several passages cover the changing ratios between night and day throughout the year. (You can read the Book of Enoch at http://www.sacred-texts.com/bib/boe/boe075.htm)

Enoch describes how the days lengthen as we move past the equal division of day and night (12 hours of daylight) at the spring equinox, telling us that at a certain point "On that day the day is longer than the night by a ninth part, and the day amounts exactly to ten parts and the night to eight parts." (Enoch 72:10) Next the ratio is 11 to 7. (Daylight lasting 14 hours and 40 minutes.) Finally at the summer solstice Enoch 72:14 says: "the day becomes double the night, and the day becomes twelve parts, and the night is shortened and becomes six parts." (Daylight is now 16 hours.)

The problem is that such a division between day and night is not possible anywhere near the latitude of Israel. Jerusalem is approximately 31.72 degrees north of the equator. As we can see on the chart of day length as a function of latitude at this web site[244]

A latitude of approximately 50 degrees north or south of the equator would be necessary for a division of the day into 16 and 8 hour sections. But this site[245] confirms that Jerusalem has an absolute maximum of 14 hours and 13 minutes between sunrise and sunset on the longest day of the year. Even the length of daylight hours specified by Enoch a full month ahead of the summer solstice is too long to be possible – at the latitude of Israel.

This means either Enoch was very specific yet very inaccurate – or that he was writing with a perspective nowhere near Israel. His description is a better match for southern Canada, somewhere between the latitude of Quebec and Winnipeg.

I have previously suggested that Enoch's 365 year lifespan is a solar reference and that the breakdown of his life into 65 years before fatherhood, and 300 years after, could be a reference to solar zenith passages and could clue us in to the latitude the site of Jerusalem was at prior to the last pole shift when the North Pole was just west of Hudson Bay.

Enoch was Noah's grandfather, and would have written about the antediluvian world before the Great Flood... but the land now called Jerusalem was at approximately 11.5 degrees North latitude before the last pole shift. Enoch apparently did not write from the perspective of Israel, even if we consider its previous position prior to the last pole shift. Where might Enoch have been located before the last pole shift that was approximately 50 degrees latitude?

There is one other very interesting possibility that comes to mind: Atlantis. If I were a college professor, writing about Atlantis would end my career. But if we're talking about pole shifts, Atlantis is the name we know for the great civilization that was destroyed in the last pole shift. And according to some researchers, Antarctica was Atlantis. Rand and Rose Flem-Ath suggest in their groundbreaking book: When the Sky Fell: In Search of Atlantis – that western or Lesser Antarctica (the part near Argentina) was a temperate land with a capital city approximately 50 degrees south of the old equator before the last pole shift. They compare the capital city of Atlantis with London, as both are capitals of world dominating seafaring empires at similar latitudes.

Could Enoch 's description of the 2 to 1 balance of day and night (a maximum of 16 hours of day and 8 hours of night) have described the summer solstice in Atlantis? It does not accurately describe the balance of day and night for Israel – but might it preserve a record from what was once the capital city of the most powerful empire before the Flood?

Enoch 77:8 describes what I believe are the seven continents from a unique perspective: "Seven great islands I saw in the sea and in the mainland: two in the mainland and five in the Great Sea."

Now if I lived prior to the last pole shift on the island of what we now call Lesser Antarctica (which is icebound to Greater or East Antarctica and displayed as a single continent on modern maps, but in reality has a large sea between the two sections of Antarctica) and this was Atlantis, the dominant world empire – I would undoubtedly place my maps with my homeland on top of the world – much as Europeans have done. We consider it natural to place Europe at the top center of a world map, and I assume anyone in Atlantis put what we call the South Pole at the top, just as the ancient Egyptians did.

So on my theoretical Atlantean map, the frozen wasteland of East Antarctica would be spread wide across the top much as Greenland is distorted on our maps (on a Mercator projection, Greenland appears as large as Africa but is really 1/13th the size.) The habitable "mainland" on my Atlantean map would be the nearby Americas – since Lesser Antarctica almost touches South America at the Straits of Magellan. Enoch's description of two continents on the mainland (North and South America) and five continents in the Great Sea makes more sense from this perspective.

From this perspective Israel would be below Atlantis, in the Atlantis map's southern hemisphere. This helps make sense of Enoch's description of the south in Enoch 77:1 "the south, because the Most High will descend there, yea, there in quite a special sense will He who is blessed forever descend." This certainly sounds like the (for Enoch, the future) incarnation of Jesus Christ in Israel.

Enoch 80:5-6 describes a future problem in "the days of sinners" when "in those days the sun shall be seen and he shall journey in the evening on the extremity of the great chariot in the west and shall shine more brightly than accords with the order of light. And many chiefs of the stars shall transgress the order (prescribed). And these shall alter their orbits and tasks, and not appear at the seasons prescribed to them." I take this to be a description of a pole shift, in which the sun is seen at what were nighttime hours and the celestial bodies appear to move out of their normal paths in the sky.

Enoch 83:3-7 "I saw in a vision how the heaven collapsed and was borne off and fell to the earth. And when it fell to the earth I saw how the earth was swallowed up in a great abyss, and mountains were suspended on mountains, and hills sank down on hills, and high trees were rent from their stems, and hurled down and sunk in the abyss... the earth: it must sink into the abyss and be destroyed with a great destruction."

This is even more clear in verses many chapters earlier in Enoch 57:2 "the pillars of the earth were moved from their place, and the sound thereof was heard from the one end of heaven to the other, in one day." Also Enoch 60:16 "the sea is masculine and strong, and according to the might of his strength he draws it back with a rein, and in like manner it is driven forward and disperses amid all the mountains of the earth."

So if anyone wonders why the Book of Enoch was not included in the Bible – I assume it is because Enoch wrote about Atlantis and a pole shift. Periodic cycles of destruction do not fit traditional interpretations of Christianity; such evidence will be swept under the rug."

In October 2017 I just stumbled on evidence that someone else reached similar conclusions on Enoch and the latitude he must have written from over a century before I did… An Australian named John Wood Beilby wrote <u>Eureka: An Elucidation of Mysteries in Nature</u> in 1883, stating that: "The authenticity of the Book of Enoch, rescued from oblivion by Bruce, the Abyssinian traveller, has been doubted, because apparently composed by two writers, one of whom must have resided south-west of the Erythrean Sea, that is, of the Persian Gulf, Red Sea, and that part of the ocean which washes the shores of Arabia, therefore the place described by him must have been south of the Abyssinian Mountains, and in or about our present definition of latitude 9° north, and longitude 36° east. He, however, or his alleged co-author, describes the longest days as containing twelve parts out of eighteen, the whole number into which the day was then divided. Now twelve are to eighteen as sixteen to twenty-four hours in our mode of computation. Ergo, the theologian reviewers hold that this Book of Enoch could not have been written in latitude 9°, but that one of its authors must have resided …within latitudes 45° to 49° north, where the longest days are from fifteen and quarter to sixteen hours. But if a point south of the Pyramids, now in latitude 9° north, was—when this Book of Enoch was written, prior to the Flood of Noah (assuming its internal evidence of being the work of Enoch or even of any ancient scribe)—in north latitude 45" to 49 degrees…" then, Beilby realized, the surface of the Earth has moved quite a bit.

Back in 1883, he suggested: "The deposit and accumulation through centuries of ice or water upon one segment of the globe in disproportionate bulk to what, under same natural laws, but varying conditions, is amassed elsewhere, and the consequent flow to and deepening of the circumjacent oceans, must produce, in time, leverage adequate to increase the tilt, and ultimately endanger the equilibrium sustaining the globe in any given position upon the plane of her orbit." Because it so strongly hints at a pole shift, I will again quote Enoch 57:2 "the pillars of the earth were moved from their place."

Returning to the remainder of an article of mine quoted extensively earlier in this book – "The History and Truth About Pole Shifts on Earth" – that article had a second section more focused on pole shift clues in Bible prophecy:

"Scientific evidence clearly proves that the planet is billions of years old. But the current surface conditions of the world – the version of the 'earth' that the Bible suggests was created just thousands of years ago – probably was created via pole shift just thousands of years ago. And because the Bible offers great detail and insight on past and future pole shifts, we should not ignore Bible history or prophecy. We should look at the Bible from the perspective of: how would ancient writers try to convey an understanding of cosmic cycles of destruction across thousands of years of dark ages? One online article (Joshua's Long Day) agrees that many 'pole shifts are recorded in the Bible, if you know how to look for them.'

In some verses the pole shift references are obvious, such as Isaiah 24:1 – 'The Lord lays the earth waste, devastates it, distorts its surface and scatters its inhabitants.' Some examples require more insight. Isaiah 45:18 tells us 'He is the God who formed the earth and made it, He established it and did not create it a waste place, but formed it to be inhabited.' This very clearly says that God 'did not create it a waste place.' Yet at the very 'beginning' in Genesis the Bible says 'The earth was formless and void.' Many scholars believe 'was' should be read 'became.' One top Bible scholar (Arthur Custance) and expert on the original Aramaic and Hebrew versions of Genesis wrote that the Bible should be properly translated to begin as: 'IN A FORMER STATE GOD PERFECTED THE HEAVENS AND EARTH; BUT THE EARTH HAD BECOME A DEVASTATED RUIN.' Could a pole shift have occurred 'in the beginning' of our present world as described by the Bible? I think that when Genesis 2:4 tells us "These are the generations of the heavens and of the earth when they were created" that we are being told of more than one (re)creation moment in history.

Pole shifts seem to be described repeatedly, from being kicked out of Eden… to the future creation of a new heavens and a new earth in Revelation. Job 9:5-6 reads 'It is God who removes the mountains; they know not how, when He overturns them in His anger; who shakes the earth out of its place, and its pillars tremble.' Hebrews 12:26-27 'Yet once more I will shake not only the earth, but also the heaven. This expression, 'Yet once more,' denotes the removing of those things which can be shaken.' The Bible even comments on its own mention of 'yet once more' to emphasize that this has happened before, and will happen again! In addition to describing pole shifts in dozens of passages, Bible prophets also use astronomical clues to describe the skies during these prophetic events and tell us when the next pole shift will occur. My last book, End Times and 2019, reviews the evidence pointing to a pole shift coming in December 2019."

…I realize that history has proven all previous attempts at date-setting wrong so far. I realize that no one has successfully calculated a date for end times events. I'm aware that Matthew 24:36, in slightly mistranslated English Bibles, says that "no one knows the hour or the day." But on the other hand, I also know that in the original Greek the verb tense in that verse references only the past; "no one has known yet" – back when this was said 2,000 years ago – would be a more accurate understanding – and that doesn't rule out the eventual calculation of end times dates.

I am torn because I realize that you, dear reader, may very well be reading this after 2019 and view the focus on this time frame in your past as misguided nonsense. But I would not be intellectually honest if I gloss over all the clues that do seem to point to December 2019. Perhaps I will be lucky enough to look back at these years from now and make an analogy to jigsaw puzzle pieces that, although their shapes fit together perfectly, did not combine to form a relevant image. You can understand the hesitation to take such pieces back apart when they fit together so perfectly. With such thoughts in mind, I have decided to include my reasoning for thinking that a pole shift could be due in December 2019, even though it is less than two years away and that timing will, hopefully, seem silly by the time you read this.

"December 21, 2019 – many clues point to it as the start of a week-long pole shift that ends with Judgment Day. The famous Mayan date of 12/21/2012 exactly seven years earlier is a noteworthy sign of an ever-worsening tribulation in which we are already experiencing problems such as earthquakes, plagues, and "wars and rumors of wars" like never before. The Mayan Pyramid of Kukulkan was built to line up with the skies of the early 21st century, symbolizing the return of their sky god to return as King over the Earth, as the Bible indicates Jesus will do at this time.

Egypt's Great Pyramid may also clue us in to this timing. Edgar "The Sleeping Prophet" Cayce hinted at this (but gave no date) when he said: "...as might be calculated from the Pyramid - there will be the beginning of the change... covered by the prophecies in the pyramid."[246] If completed to include its missing top, the Great Pyramid would be 5780 inches high. Several British Egyptologists believed that the inch is an ancient unit; that the Great Pyramid's measurements in inches equal years in a prophetic timeline; and that the entire pyramid represents the world. 5780 inches in height may represent completing the unfinished pyramid, and "completion of the world." Hebrew calendar year 5780 corresponds with December 2019. Is this measurement an Egyptian prophecy of the end or "completion" of the world as we know it in 2019?

The Bible has many clues pointing us to this date. It tells us in Haggai 2:20-21 that God will start to "shake the heavens and the earth" on the 24th of the month of Kislev, the night before Hanukkah begins on Kislev 25. In 2019, Hanukkah starts on December 22, so if this is the year God shakes the heavens and the earth, He starts on December 21. Matthew 24:13-20 tells us we must take flight immediately when we know the end is coming. Verse 20 says "pray that your flight will not be in the winter, or on a Sabbath." December 21, 2019 is the beginning of winter, and is a Saturday (Sabbath). And it is exactly seven years of tribulation after the date the Maya warned us about.

But the most impressive clues involve "forensic astronomy" – using descriptions of the sky as clues to timing future events like the pole shift. Many prophets from Nostradamus to Isaiah describe the night sky at the time of their future visions. Amazing matches occur between their astronomical signs in the sky and the actual positions of heavenly bodies during the week from December 21-28, 2019. This week may also correspond to the final seven day warning God gave Noah in Genesis 7:4. The sun, moon, and planets represent the characters in a wedding by moving through the sky in a way that corresponds to acting out all the major steps of an ancient Jewish week-long wedding ceremony.

As I wrote in a previous book on the Antichrist: "He has placed a tent for the sun, which is as a bridegroom coming out of his chamber." (Psalm 19:5) This very clearly tells us that the sun coming out of its appointed set path/place is as the bridegroom exiting the bridal chamber at the end of the wedding.

I know of one biblical bride who became Queen by marrying the King on a known date. Esther married the Persian King Ahasuerus (Xerxes) on the first day of the month of Tevet, as hinted at in Esther 2:16. Christ is the King of Kings, and if I am correct, Christ's bride will reign as Queen on the first of Tevet as well, which is now the seventh day of Hanukkah, and falls on December 28 in the year 2019."[247]

Some biblical debunkers comment on the story of Esther and her uncle Mordecai and suggest they never existed; that these are just Judaized versions of the names Ishtar and Marduk, and that the Jewish holiday of Purim, intended as an allegory to teach certain truths, is really just a modified version of an ancient Persian and Babylonian holiday celebrating the victory of Marduk and Ishtar over their celestial enemies. From our perspective in drawing an analogy to events in the sky – including a war in heaven and a celestial wedding ceremony on the seventh day of Hanukkah – the fact that Marduk's victory over Tiamat at Nibiru may have occurred on the same date thousands of years ago only supports my theory.

The sudden arrival of a galactic superwave, with radiation in all its forms – including a gravity wave, bright blue light, and electromagnetic currents – also supports these ideas... Legends and symbolism on the heavenly wedding and Solomon's "temple at Jerusalem indicate that two forms of natural energy were involved, one terrestrial, the other from the atmosphere... a strong hint that one party in the sacred marriage performed at the Temple was electrical current from the atmosphere. The other partner in this union of opposite elements [heaven and earth] was clearly the magnetic current of the earth..."[248] A similar quote in another book says: "The marriage of heaven and earth took place at the Pyramid as a union between the terrestrial current, accumulated in its rocky mass, and the divine spark of celestial fire."[249] Perhaps there is a cosmic recharging of the planetary dynamo during a cosmic wedding and war and a terrestrial cataclysm.

First the groom goes to his father's house – this is Jesus (our winter solstice sun-god) approaching the galactic center through precession, which He has finally reached in recent years. The earth (the bride) must circle the altar and groom seven times before the ceremony begins. This could be the seven years from 12/21/2012 to 12/21/2019, during which the earth orbits the sun seven times. The bride must also make herself new garments and take a ritual bath – this could be the earth's oceans sloshing out of their basins during a pole shift, with the new surface orientation being the new outer garments.

One part of the wedding ceremony gives the groom eight minutes in a private room to relax with the bride and give her a ring. On 12/26/2019, there is a total solar eclipse visible from the Middle East. Any such eclipse lasts less than 8 minutes and we on earth (the bride) can see/receive the "ring of fire" visible around the edge of the moon. The solar eclipse occurs just two days before what might be the cataclysmic day of the most extreme pole shifting on December 28, which fits well with Joel 2:31 "The sun will be turned into darkness and the moon into blood before the great and awesome day of the Lord comes." This could also be the same as the Sixth Seal in Revelation 6:12-14 "when He broke the Sixth Seal, and there was a great earthquake, and the sun became as black as sackcloth made of hair, and the whole moon became like blood; and the stars of the sky fell to the earth, as a fig tree casts its unripe figs when shaken by a great wind. The sky was split apart like a scroll when it is rolled up, and every mountain and island were moved out of their places."

The wedding couple retreats to a honeymoon room for the rest of the week until the best man knocks and hovers at the door and waits for the groom to come out with his wife. On December 28, 2019 Jupiter appears to hover at the edge of the sun, repeatedly disappearing and reappearing behind the sun's corona. This is the final day of the wedding, and possibly of a pole shift – and if it happens this way we may as well call it Judgment Day.

I just learned something now about Islam's Tariq Star that appears at the end of days which is relevant here. I already assumed that this "star" is really the newly visible bright blue light of the "active" or "Seyfert phase" of the galactic center. I just learned that in Arabic, this "star" is also known as "the knocker" from the root wood TARQ which means "knocking loudly." In the Qu'ran, Surah 86 begins: "O, by the heaven and the knocker... the piercing star... to resurrect man, God is able on the day when secrets are revealed..." So the Tariq Star is an important heavenly body present at the end of days, and it is at that point knocking at the door... This may be related to the Bible's Revelation 3:20-22: "Behold, I stand at the door and knock; if anyone hears my voice and opens the door, I will come in to him and will dine with him, and he with Me. He who overcomes, I will grant to him to sit down with Me on My throne, as I also overcame and sat down with My Father on His throne. He who has an ear, let him hear what the spirit says..." Knowing the importance of the active cycles of the galactic center, are we sufficiently initiated to have an ear to hear, and to understand that this knocking at the door is symbolic of a wedding in the sky and of a catastrophic pole shift on the ground?

I also recently discovered a comment made long ago (1955) by Adam Barber, who wrote in <u>The Coming Disaster Worse Than the H-Bomb</u>: "...the next shift causing a great flood will occur on December 21st or June 21st of any year." I am reminded of "the Roman philosopher Seneca, discussing the Babylonian cosmology of Berossos... that the future flood will take place when [a certain] conjunction takes place in Capricorn. For the former is the constellation of the summer solstice, the latter of the winter solstice; they are the decisive signs."[250] Barber apparently felt that the days near the winter solstice – and to a lesser extent, the summer solstice – are the most dangerous times of year for a potential pole shift to start. Just how many "years from now I cannot ascertain or calculate, but from the astronomical data at hand I am certain it will be very soon." I think Barber expected the next pole shift in the late 20th century or early 21st century – then again, I don't know what "astronomical data" he based it on. (Like Emil Sepic, who wrote "The Imminent Shift of the Earth's Axis," Barber believed the Earth makes a second, minor orbit causing an irregularity in our annual rotation of the sun, corrected occasionally by a pole shift.)

In End Times and 2019, I based my pole shift start date of December 21, 2019 mostly on Bible prophecy, backed up by many astronomical matches to the prophets' visions of the night skies, to other calendric prophecies with Hebrew calendar dates given, and also to Mayan and Egyptian calendars, myths, and other clues. Since writing that book, I have come across more clues in mythology indicating that the last pole shift, half a precession cycle back around 12,900 years ago – occurred on the summer solstice – which further emphasizes the importance of a solstice date for future prophecy. I also finally realized that the summer solstice on one side of the equator is the winter solstice on the other side. It may not matter much which solstice certain information points to – because both are right for one half of the world. The importance of summer or winter timing depends on where a "myth" developed and how much the crust moved...

And even with dozens of clues pointing to December 2019, nothing makes that timing a certainty – just a timeframe worth paying attention to. I cannot emphasize enough that although there is significant evidence suggesting a specific time for the next pole shift, and that I value End Times and 2019 as an informative book no matter what ends up happening – there is no certainty about it. I followed clues to a conclusion, but it is only a theory. I don't want to dwell too much on the idea that the next pole shift is due on any particular date as if we have enough evidence to be one hundred per cent certain.

Because there is evidence pointing to December 21-28, 2019 I think it merits our attention. It may be a very eventful week. Many years ago I thought it would mark the end of a seven year tribulation. Maybe it will mark the beginning. And maybe nothing will happen at the end of 2019. Maybe, like someone staring at a Rorschach ink blot until they see a pattern – maybe I found a meaningless pattern in the prophetic and astronomical correlations I saw. There are other clues pointing to years slightly farther into our future for a coming pole shift. Let's just acknowledge that a great deal of evidence points to an upcoming pole shift "soon."

Terrence McKenna also comes to the conclusion that a solar eclipse, in conjunction with the winter solstice sun's arrival near the galactic center, could mark the end of time for our present world age. "The brothers Dennis and Terence McKenna made a computer analysis of the Chinese oracle system of the I Ching (Book of Changes), which dates to circa 1,000 BC. The I Ching consists of 64 arrangements of 6-line hexagrams (kua). The permutations of the lines express universal archetypes which illustrate fate. The McKennas discovered that the I Ching also represents a highly accurate lunar calendar of 384 days, divided into 13 months. The calendar was lost in a mass book-burning in the 3rd century BC, after which it was replaced with a less accurate 360-day calendar. Further research revealed that the I Ching calendar ends in the year 2012 --- within a year of the end of the South American Eagle Bowl Calendar!"[251]

Is there a reason that so many calendars, myths, monuments, legends, and prophecies seem to point to the seven year period from 2012-2016-2019? "The McKenna's offered an explanation for this synchronicity in their book, The Invisible Landscape:

"The second approach to a search for possible dates of future concresence is more subtle and takes account of the precession of the equinoxes. Because of the precession, the solstice and the equinoctal nodes precess or move backwards against the background of the fixed stars which comprise the zodiac. In a 26,000-year zodiacal great year, the solstice and equinoctal nodes move around the entire zodiac. It is a coincidence then that in our own time, the winter solstice is placed in the constellation Sagittarius, only about 3º from the galactic center which, also coincidentally, is within 2º of the ecliptic. Because the winter solstice node is precessing, it is moving closer and closer to the point on the ecliptic where it will eclipse the galactic center. This will occur sometime within the next 200 [this might be a typo for 20] years. It is difficult to be more accurate, since the term "galactic center" is ambiguous. A degree covers a large area, and the galaxy may be presumed to have a gravitational center, a radio center, and a spatial center. Nevertheless, we suggest that the transition from one zodiacal era of approximately 2,200 years duration to the next may be hinged on the conjunction of the solstice node and the galactic center. It is useful to examine winter solstices on which solar eclipses will occur over the next 200 years, during which the earth's solstice node will be slowly transiting the area of the galactic center. The eclipse of the galactic center by the solstice sun, which is itself in eclipse relation to the earth, might be an event unusual enough to signal an onset of concrescence..."

There is a winter solstice approximately every 12/21-22 and a solar eclipse 12/25-26 in 2019. Is McKenna suggesting that these astronomical alignments could cause, or at least mark the timing of, world-ending events? McKenna also wrote: "...History is the shock wave of eschatology... It is the great, great adventure of becoming, and we are very privileged to be in this final ticking out of the last seconds of the third act."

I detail the similarities between the ancient prophets' visions of the night skies with the stages of a week-long Jewish wedding ceremony and the actual movements of the sun, moon, and planets in December 2019 for six pages in End Times and 2019. But despite the impressive correlations, it doesn't prove my timeline for end times events. Many other clues pointed to June 2016, halfway through the final seven years of my timeline, as the point when the Antichrist would seize more dictatorial power and reveal his intentions to the world.

In Antichrist 2016-2019, I suggested that Barack Obama might seize dictatorial power and attempt to remain in office long after 2016. As it turns out, Recep Erdogan seized more dictatorial control over Turkey in July 2016. Donald Trump came out of nowhere (politically, he held no previous office) and was suddenly catapulted to the most powerful political office in the world in 2016. Could one of them eventually prove to be the Antichrist described in end times Bible prophecy? It's possible, but as of early 2018 – it seems increasingly unlikely that everything will finish up by the end of next year.

Nostradamus describes an Islamic Antichrist, and specifically describes a leader from Turkey and rule from Constantinople/Istanbul. But as I explain in a more recent book (Nostradamus and the Islamic Invasion of Europe) his prophecies of a gradually intensifying 27 year conflict between the Islamic world and the West may very well cover the years from 2001-2028. Nostradamus also describes a pole shift, coming at the end of WWIII, but this seems more likely to be in 2028 or 2029, based on his prophecies – and not in

December 2019. The more I write, and the more research I do, the more I realize that different sources of information point to different years regarding the likely timing of the next pole shift.

One aspect of Bible prophecy also suggests that everything happens by about 2028 at the latest, if the generation that saw Israel re-created as a nation in 1948 is supposed to see the fulfilment of all end times prophecies, and the longest mention of a generation is 70 or 80 years... Nostradamus suggests a crucial period from 2001 to 2028 then wrote of almost nothing beyond the beginning of the 21st century... The Mayan Long Count Calendar ended in 2012, but the Dresden Codex of their Popul Vuh creation story shows a world flood during an astronomical configuration which is in effect seven years later in December 2019... The astronomical symbolism of their ballcourt game also points to the game ending when the sun-ball crosses through the hoop on December 21, 2019. Their pyramid of Kukulkan at Chichen Itza has been called "a precessional alarm clock set for the twenty-first century" but this doesn't yield a specific date. Scott Creighton sees the Egyptian "pyramids at Giza as a monumental 'alarm clock'" between 2000 and 2025.

Even a 1982 study on "Precognitive and Prophetic Visions in Near-Death Experiences" noted: "There is, first of all, a sense of having total knowledge, but specifically one is aware of seeing the entirety of the earth's evolution and history, from the beginning to the end of time. The future scenario, however, is usually of short duration, seldom extending much beyond the beginning of the twenty-first century." But none of this conclusively proves a specific date as the right answer. It should make us pay attention and at least consider the possibility that a catastrophic pole shift could bring on the end of the world as we know it very soon, in the early 21st century, when both the Bible the Maya and Nostradamus and other prophetic sources seem to expect it – but perhaps a pole shift won't happen again for a very long time. We have many hints, but we can never be sure about the exact timing.

While there may be uncertainty in timing, there is no uncertainty that the ancient writers attempting to warn us did have catastrophic pole shifts in mind. Consider these potential references to crustal displacements scattered throughout the Bible:

1 Samuel 2:8 "For the pillars of the earth are the LORD'S, and He set the world on them."

Job 9:5-9 "It is God who removes the mountains, they know not how, when He overturns them in His anger; Who shakes the earth out of its place, and its pillars tremble; Who commands the sun not to shine, and sets a seal upon the stars; Who alone stretches out the heavens and tramples down the waves of the sea; Who makes the Bear, Orion, and the Pleiades?" (In this one passage the earth is shaken out of its place, its pillars/axis shake, we have ocean waves, a darkened sun, and a mention of three constellations in the sky which may have shown signs of celestial activity in the past.)

Psalm 19:4-5 "He has placed a tent for the sun, which is as a bridegroom coming out of his chamber." (I believe the sun has a role in a celestial wedding, when the sun, moon, and planets act out all major steps of an ancient Jewish wedding ceremony from December 21-28, 2019 – and that this is a time frame worthy of attention regarding our planet's next catastrophic pole shift.)

Ecclesiastes 1:9-11 "There is nothing new under the sun. Is there anything new of which one might say, 'See this, it is new?' Already it has existed for ages which were before us. There is no remembrance of earlier things; and of the later things which will occur, there will be no remembrance among those who will come later still."

Isaiah 2:19 "Men will go into the caves of the rocks and into the holes of the ground before the terror of the Lord and the splendor of His majesty, when He arises to make the earth tremble."

Isaiah 13:10-13 "For the stars of heaven and their constellations will not flash forth their light; the sun will be dark when it rises and the moon will not shed its light. Thus I will punish the world for its evil ...I will make the heavens tremble, and the earth will be shaken from its place at the fury of the Lord of hosts in the day of His burning anger." (On December 26, 2019 there will be a total solar eclipse visible from Jerusalem at sunrise – the sun will be dark when it rises – and Israelis and Jordanians will see the sun's corona, the "diamond ring effect" during the right part of the celestial wedding taking place in the sky that week!)

Isaiah 22:25 "In that day, declares the Lord of hosts, 'the peg driven in a firm place will give way…" (Could this be "the North Nail" that so many cultures use to name the North Pole?)

Isaiah 24:1 "The Lord lays the earth waste, devastates it, distorts its surface and scatters its inhabitants."

Isaiah 24:18-20 "The foundations of the earth shake. The earth is broken asunder, the earth is split through, the earth is shaken violently. The earth reels to and fro like a drunkard, and it totters like a shack. For its transgression is heavy upon it, and it will fall, never to rise again."

Isaiah 34:4 "The sky will be rolled up like a scroll."

Isaiah 65:17 "Behold, I create a new heavens and a new earth; and the former things will not be remembered."

Jeremiah 4:22-28 "I looked on the earth, and behold, it was formless and void; and to the heavens, and they had no light. I looked on the mountains, and behold, they were quaking, and all the hills moved to and fro. I looked, and behold, there was no man, and all the birds of the heavens had fled. I looked, and behold, the fruitful land was a wilderness, and all its cities were pulled down before the Lord, before His fierce anger. For thus says the Lord, 'The whole land shall be a desolation, yet I will not execute a complete destruction. For this the earth shall mourn and the heavens above be dark, because I have spoken, I have purposed, and I will not change My mind, nor will I turn from it.'" (Note this cannot be describing an original time of creation – man exists and cities are destroyed – it seems like a future event.)

Daniel 4:10-11 "Now these were the visions in my mind as I lay on my bed: I was looking, and behold, there was a tree in the midst of the earth and its height was great. The tree grew large and became strong and its height reached up to the sky, and it was visible to the end of the whole earth."

Daniel 4:23 "The king saw an angelic watcher, a holy one, descending from heaven saying, 'Chop down the tree and destroy it; yet leave the stump...'" (To me, Daniel seems to describe a destroyed axis.)

2 Esdras 3:6, (Catholic Bibles only) when Ezra speaks of God bringing Adam into a garden in Eden that had already existed before "the earth" was created: "And you led him into the garden that your right hand planted before the earth appeared." That garden must have been started on a previous version of the planet's surface conditions, if it had been "planted before the [new configuration of the] earth appeared."

2 Esdras 3:9-18 (Catholic Bibles only) "But again, in its time you brought the flood upon the inhabitants of the world and destroyed them…. You bent down the heavens and shook the earth, and moved the world, and caused its depths to tremble…. Your glory passed through the four gates of fire and earthquake and wind and ice."

2 Esdras 2:4-8 tells us "The sun will suddenly start shining at night, and... the movement of the stars will be changed."

1 Thessalonians 5:4 "But you, brethren, are not in darkness, that the day would overtake you like a thief." This refers to the idea that Judgment Day comes like a thief in the night – a surprise to those who are (spiritually/intellectually) asleep – uninitiated in the mysteries. It really refers to the practice of the Temple's High Priest to make his rounds overnight to make sure the other temple priests keep the sacred fire lit and keep the temple safe. If they fall asleep on their watch, he would set their robes on fire. If they were awake, he appeared as a friendly visitor at an expected, appointed time. Many Christians believe that Jesus will return like our High Priest, like a thief in the night. They misunderstand that this only happens at an unexpected time if you fall asleep on your appointed watch. If you are awake, you know when He is coming.

Hebrews 12:26-27 "'Yet once more I will shake not only the earth, but also the heaven.' This expression, 'Yet once more,' denotes the removing of these things which can be shaken..." (This verse emphasizes how a pole shift has happened before and will happen again.)

2 Peter 2:5 God "did not spare the ancient world."

2 Peter 3:6-8 "the world at that time was destroyed, being flooded with water. But by His word the present heavens and earth are being reserved for fire."

Jude 6 may be describing stars that fell below the horizon in a pole shift: "And angels who did not keep their own domain, but abandoned their proper abode, He has kept in eternal bonds under darkness."

Matthew 24:21-22 "For then there will be a Great Tribulation, such as has not occurred since the beginning of the world until now, nor ever will. [Pole Shifts happen at the beginning and end of worlds…] Unless those days had been cut short, no life would have been saved; but for the sake of the elect those days will be cut short." This could mean that for the sake of the surviving few, the number of days of tribulation is kept to a minimum. It could also mean that the days themselves are shorter than 24 hours. This presumably would require the sun to appear to traverse the sky faster than usual, which could happen if the surface of the planet were rotating abnormally below it.

Matthew 24:29-31 "immediately after the tribulation of those days the sun will be darkened, and the moon will not give its light, and the stars will fall from the sky…"

Revelation 3:3 "If you do not wake up, I will come like a thief, and you will not know at what hour I will come." As with 1 Thessalonians 5:4, it is suggested here that if we are awake and initiated with knowledge, we will know the timing of end times events.

Revelation 6:12-14 "I looked when He broke the sixth seal, and there was a great earthquake; and the sun became black as sackcloth made of hair, and the whole moon became like blood; and the stars of the sky fell to the earth, as a fig casts its unripe figs when shaken by a great wind. The sky was split apart like a scroll when it is rolled up, and every mountain and island were moved out of their places."

Revelation 9:2-3 "He opened the bottomless pit, and smoke went up out of the pit, like the smoke of a great furnace; and the sun and the air were darkened by the smoke of the pit. Then out of the smoke came locusts upon the earth, and power was given them, as the scorpions of the earth have power." (The black hole at the galactic center is a bottomless pit that occasionally spews out radiation and great clouds of dust. When a superwave reaches our solar system, the cosmic dust pushes in and overcomes the solar wind that normally keeps the solar system relatively free of dust. As these dust particles fill our region of space, visible sunlight would be blocked out. Dust particles would constantly be falling into our atmosphere by the trillions, and anything that reaches the surface at a size near a grain of sand but at extremely high speed would certainly sting any person or animal it hits. The problem would appear to arrive from the new light at the galactic center, near the stinger of Scorpio.)

Revelation 9:5-6 continues to say that these "torment for five months; and their torment was like the torment of a scorpion when it stings a man. And in those days men will seek death and will not find it; they will long to die, and death flees from them."

Revelation 11:3 the "two witnesses" are "clothed in sackcloth." I can't rule out divine intervention and the reappearance of human end times witnesses like Elijah, with someone like Moses or Enoch or the Apostle John with him – I have heard arguments in favor of all of them. But perhaps the two witnesses, symbolized astronomically, are the sun and the moon. Could the incoming cosmic dust be what clothes them in sackcloth?

Revelation 11:12 After the death of the two witnesses, they disappear out of sight and go "up into heaven in the cloud." Not just a cloud, but the cloud, as if there is one very prominent cloud.

Revelation 16:18-20 "There was a great earthquake, such as there had not been since man came to be upon the earth... And every island fled away, and the mountains were not found."

Revelation 21:1 "Then I saw a new heaven and a new earth, for the first heaven and the first earth had passed away."

The above quotes from the Bible are just a tiny fraction of the ones I could use in support of the idea that the Bible describes past and future pole shifts. In

End Times and 2019 I have a much more extensive list of biblical "pole shift" quotes from page 219 to page 255.

As Gerald Massey concluded in his books over a century ago: "The fulfillment of scripture was the completion of astronomical cycles."[252] I would go a step further and suggest that there is a reason why our ancient ancestors were so obsessed with astronomical cycles. They knew such cycles were linked to devastating pole shifts on earth, and they used astronomical cycles for both shorter term calendar needs, and to keep track of world ages and catastrophes over extremely long periods of time – possibly with the hope that if knowledge of the cycles of destruction could be preserved, a future civilization like ours could prepare better and avoid being knocked back to the stone age. I believe they made sure an initiated brotherhood, "stewards of the mysteries of God" (1 Corinthians 4:1) a secret society of elite, in-the-know people maintained the ancient wisdom into modern times. If so, have they secretly prepared?

"Hebrew stories are influenced by the Egyptian idea that celestial wisdom is to be a secret understanding between the initiated priesthood 'who were able to read astronomical allegories in which deeper secrets, not granted to the common herd, lay concealed. We presume that these people were once called the Followers of Horus.'"[253] Acts 7:22 tells us that "Moses was educated in all the learning of the Egyptians." Jesus also spent years in Egypt, (Matthew 2:14 "Out of Egypt I called My Son.") and of course even in Israel the Jewish people were under the heavy cultural influence of their powerful neighbors and former masters in Egypt. It should be no surprise that Christ often addressed himself only to those who have ears to hear – those who are initiated into the ancient wisdom of the cosmos, to understand that much of what is described using people on the ground truly pertains to the movement of heavenly bodies. Especially when the events described in the Bible are not possible literally – I think they are meant to describe astronomical events, cycles, and the pole shifts that occur at the end of such cycles.

Many stories in the Bible, like so many other ancient stories from around the world, often describe characters and events that are not literally possible. Some would suggest that myths and stories and implausible Bible tales should have their events be discarded out of hand. They suggest that in such stories only the outcomes mattered, and the "once upon a time" beginnings and opposite-of-reality adventures in the middle are just nonsense to allow the strange end results. I suggest that when biblical events cannot possibly be taken literally, we should view them astronomically. As the authors of Hamlet's Mill say it: "There are many events, described with appropriate terrestrial imagery, that do not, however, happen on earth."[254] Samson is one of these implausible, astronomical characters.

In Judges 14:5-6 Samson (whose name means "belonging to the sun") tears a young lion in half with his bare hands. Should we take this at face value? Or should the young lion be viewed as the beginning of the Age of Leo, when the sun started rising in Leo at the spring equinox? In Judges 15:15 Samson slayed a thousand men with the jawbone of an ass, then threw it, then declared his thirst to God, who then made water come out of the jaw. These "facts" sound like nonsense, until we compare them to the Babylonian story of their god Marduk, who uses the Hyades star cluster – "the watery jaw of Taurus" – as a boomerang type of weapon to kill a group of heavenly monsters.

If during the Taurid meteor showers, one or two huge meteors hit the Indian Ocean about 5,000 years ago and caused the Great Flood, then the events might get linked to Taurus in ancient stories written by the survivors. This possibility seems to be supported by Jewish stories in the Talmud that describe God removing two stars from the Pleiades and reversing the order of the universe at the time of Noah's Flood. The "upper waters" of the sky poured through the holes of these missing stars in the Pleiades, until God replaced them and plugged the holes with stars from the Bear, Ursa Major. These constellations became associated with destruction – Achilles even put Ursa Major and the Pleiades on his shield when he led Greek armies in the destruction of Troy...)

Samson's wife Delilah's name loosely translates into the words "hers" and "water pitcher" and may mean "she of Aquarius." If so, we now have three of the four cardinal zodiac signs mentioned in Samson's story, the Ages most likely to be in effect at the time of recurring cosmic catastrophes, based on the historical analysis and especially on the charts of Randall Carlson.

In <u>End Times and 2019</u>, I suggest that all the major steps of a week-long, ancient Hebrew wedding ceremony will be acted out in the sky as we enter the age of Aquarius. Perhaps we should be on watch, as many clues in Bible prophecy suggest that we might have been warned of a pole shift from December 21-28, 2019. I realize my correlation to alignments in the skies may prove meaningless – I hope it passes uneventfully and you are reading this after 2019 – so without dwelling on the exact timing – the key point here is the correlation to a wedding in the sky as a marker or clue to future events. The Egyptians wrote about it. Plato wrote about it. Christian eschatology ends with Christ coming back for His bride.

In Judges 14, Samson arranges a week long feast that appears to be a wedding party. With the idea in mind that these periodic pole shift catastrophes are triggered by the arrival of a galactic superwave every half precession cycle – remember that Samson represents the sun. Judges 14:6 "The Spirit of the Lord came upon him mightily, so that he tore him..." (Energy from the galactic center energized the sun in the Age of Leo, and Samson tore the young lion in half with his bare hands – he destroyed the Age of Leo as it was no more than halfway through.) In Judges 14:7-8, a woman looks good to him, so he comes to take her, and notices the corpse of the lion filled with bees and honey, which he eats, and gives to his parents to eat. This is also symbolic and impossible.

Looking up the facts on bees and honey production, we will see that a newly established hive rarely produces any honey the first year. If climate is very favorable and it does produce honey the first year, it still requires waiting from the start of hive-building in spring, until some honey is made in the fall. Making the wax honeycombs first takes a lot of work! Even a hive already established in a previous year needs at least 5-6 weeks to produce significant honey in pre-existing wax honeycombs. So when Samson kills a lion, and two sentences later there is enough honey in it to share with several people, this must be symbolic. For the lion to rot and dry out and have bees build a hive and produce honey – there hasn't been enough time passing in two sentences.

This is hinted at when Samson asks the groomsmen a riddle about it in Judges 14:12 "Out of the eater came something to eat, and out of the strong came something sweet." As with the Sphinx incorporating zodiac signs and asking a riddle for us to solve, we are the ones being questioned by Samson here, to decipher the meaning – that the Golden Age (sweet honey) was in the Age of Leo. The groomsmen cannot decipher the riddle, and threaten Delilah, and coerce her to find out the answer and tell them. On the seventh and final day of their week long wedding feast, Delilah's tearful begging finally got Samson to tell her, and she told the groomsmen. In verse 18 Samson replies: "If you had not plowed with my heifer, you would not have found out my riddle." This tells us that they got the answer from Delilah, "she of Aquarius."

Did our ancestors understand one pole shift event by looking back to evidence of the previous catastrophe in the Age of Aquarius (now 25,900 years ago) much like we look back at the previous catastrophe in the Age of Leo to understand what might happen in our near future? If I am interpreting this correctly, then Samson's killing spree on the last day of the wedding feast corresponded to the to the last day of a week-long wedding between heaven and earth, the end of the golden age of Leo on the final day of the pole shift. The next time such a wedding and pole shift occur might as well be called Judgment Day, and I think it will be early in the Age of Aquarius.

Is there other pole shift "evidence" in Samson's wild stories? In Judges 15:8 Samson "struck them ruthlessly with a great slaughter; and he went down and lived in the cleft of the rock of Etam." This would have occurred as the summer solstice sun was at the Great Cleft of the Milky Way by the galactic center. (Many Mayan stories focus on the winter solstice sun arriving at the dark cleft of the Milky Way at out next catastrophe.) In Judges 16:3 Samson was expected to rise at dawn (like the sun normally does) but he didn't get up until midnight (maybe the sun was already not following its normal schedule through the sky, because a pole shift had begun?) and he pulled out the doors of the Gaza city gate "and the two posts and pulled them up." Might the two posts being removed refer to a change in the axis of the Earth's rotation?

When Delilah finally shaves Samson's hair and his strength and the Spirit of the Lord depart from him (some clues from the past suggest that the sun's corona is lit from electrical discharge onto its surface, and that it will go out for three days of darkness during a null period when it isn't receiving electrical energy from the galactic center) the Philistines finally seize him and gouge out his eyes (another hint at darkness, and the sun and moon being dark...) First they force him to work (Judges 16:21) as "a grinder in the prison." What have we learned about grinding at the mill, in regard to the long cycles of astronomical rotation, and the eventual destruction of the mill – in regard to a pole shift? The Philistines eventually bind Samson to the two pillars holding up their great house. This is an odd architectural design, to have two central pillars just a few feet apart – unless the two pillars really refer to the two poles of the earth's axis. Judges 16:30 says "And he bent with all his might so that the house fell on the lords and all the people who were in it. So the dead whom he killed at his death were more than those whom he killed in his life." The final day of a pole shift has the most fatalities.

The New World Translation of the Bible refers to these Philistine lords as "axis lords." (I find it interesting that in Judges 16:5 the five Philistine axis lords offer Delilah 1100 silver pieces each – 5500 total - to betray Samson. Hugh A. Brown repeatedly mentions a pole shift of "5,500 miles" when the old poles move approximately 75-80 degrees latitude to end up near the new equator.) The Hebrew word used for lords (only of these Philistines or axis lords, and nowhere else in the Bible) is "seranim" which shares the same consonants as the word "axle" as found in 1 Kings 7:30... That chapter in the Book of Kings is all about Solomon and Hiram setting up the Temple (which might represent the earth) and its many measurements, including those of the "two pillars" mentioned in verses 7:41-42. If Samson represents the sun at the beginning of the Age of Leo, and his enemies were "axis lords" and he pushed against "two pillars" and "bent" them until everything collapsed – I think we could be looking at an odd description of the last pole shift around 12,960 years ago. The winter solstice sun would also have been seen near Orion, the only blind constellation figure.

Samson's story is repeated around the world under different names. There are many mythical characters forced to grind at the mill. Many characters have one or both eyes removed. Samson's destruction of two pillars at Gaza sounds a lot like Hercules destruction of two pillars at Gadiz. It sounds a lot like Japan's Susanowo, whose hair was tied to the palace rafters before he pulls down the pillars of the palace. (Earlier in his story, Susanowo seems to be involved in potential pole shift earth changes... he arranges the "Drawing of the Lands" involving the creation of new lands and giving the Japanese islands the shapes they have today.) Samson also sounds like the mythical Maori hero Whakatau in New Zealand, who pulled ropes tied to the posts of the great house and brought it down. It sounds like the story of Zipacna in the Mayan Popul Vuh creation epic, when he brings down the giant ridgepole of the house and kills everyone inside it.

There is another Christian theme with roots around the world: Christ's death on the cross. Christians would like to believe this story is unique, but evidence shows that in many details, Jesus Christ's story is an updated version of the story of Krishna, Mithras, Horus, Quetzalcoatl, Dionysus, and many other sun-gods. Many are born to a virgin around the winter solstice, their birth heralded in advance by a star. Many had someone with a name like Herod or Herut out to kill them as a baby. Many were baptized in water by someone who was later beheaded. Many were tempted in the desert by Set or Satan, had twelve disciples and a last supper, cured blindness and leprosy, brought the dead back to life, and had titles like "King of Kings," "Lord of Lords," "Redeemer," "Savior," "Anointed One," and "Son of God." (If interested in all the details, read Kersey Graves' The World's Sixteen Crucified Saviors or Suns of God by Achyra S.)

Various sun gods have died, descended into hell or the underworld, and were resurrected three days after being sacrificed to save humanity through a very temporary death on a cross, or the crossing of the four roads, or the crossing point of the Milky Way and the ecliptic. This is the time and place at which they ascend to their father, the highest god, and receive great power and kingship over the earth.

Astronomically, there are several key similarities. The throne of the Supreme God is at the galactic center, at the crossing point of the Milky Way and the path of the ecliptic. The Son of God represents the Sun. The son/sun temporarily dies at this crossing point/throne, becomes one with his father, and receives great power to destroy and remake the Earth, purifying the world by saving a remnant of the good, and wiping out the wicked majority. And the Sumerian word for crossing point is Nibiru.

Nibiru – Fiction and Fact

If you started reading this book because of an interest in Nibiru you have been very patient. Nibiru may very well be an important cause of pole shifts, or at least trigger the Earth into a pole shift when other forces are imbalanced and near the point of catastrophe already. But Nibiru is not what most people have been led to believe, so let's begin by considering what most people seem to think about Nibiru...

The idea I most often encounter on the internet is that Nibiru is either a rogue planet that will pass through our solar system, or a planet that belongs to our solar system but has such a unique and elongated orbit that it is brought towards the inner solar system (causing cataclysmic problems on the Earth) every 3,600 years. Nibiru is often associated with "Wormwood," "The Destroyer," and especially "Planet X" – the idea that a yet undiscovered planet beyond the orbit of Neptune would explain the orbital irregularities of the outer planets.

To a lesser degree, Nibiru is also described as a hidden binary companion of our sun, a red dwarf or dark star orbiting our sun, dimly lit and far enough away that no one has conclusively proven its existence. Both ideas are confused with and apparently interchangeable with Nemesis, Wormwood, and The Destroyer. As a small star or a large planet, Nibiru is believed to have several of its own planets or moons orbiting in its system. It is usually depicted as reddish in color, and alleged photographs of it often show these orbiting companions. It is sometimes depicted as having hornlike or wing-like protrusions trailing behind it.

Nibiru is often cited as a likely cause of upcoming catastrophe in our near future. Due to the popular idea of a 3,600 year orbit that repeatedly brings it near Earth, (or a 3,657 year orbit which allegedly takes Nibiru around our sun and it's dark binary companion, as believed by those at poleshift.ning.com) Nibiru has also been mentioned as the cause of past events like the three hours of darkness at the Crucifixion, the parting of the Red Sea, and especially Noah's Flood. Its close approach, many believe, is the cause of various earthquakes and floods in recent years. Millions of people think Nibiru can be seen from Antarctica right now, and that NASA and other government agencies have orchestrated a massive conspiracy to cover up the existence of Nibiru.

The idea of a rogue planet like Nibiru first took the spotlight around 1976, when Zecharia Sitchin started publishing books like The 12th Planet, the first in a series of six books about an alien race called the Anunnaki from the planet Nibiru and their coming to Earth to mine gold, genetically engineer humans from apes and rule over early humanity as gods. Sitchin claimed to be a scholar and expert on ancient Sumeria and other related topics, and that his mastery of several related languages and topics allowed him, and only him, to accurately understand ancient Sumerian writings and artwork. He claimed that from his unique understanding of their culture, he was the first to accurately understand their historical stories about the Anunnaki coming from the planet Nibiru.

In actual Sumerian texts, the Anunnaki are divine beings relegated solely to the underworld; their heavenly counterparts are the Igigi. They are not associated with Nibiru at all. At the time of the Flood, all the Anunnaki were said to do in Sumerian texts was raise their torches and light up the land, and cry over the destruction. These "gods" were "terror stricken at the deluge… the gods cowered like dogs (and) crouched in distress." The goddess Ishtar cried over the destruction of most of humanity, and "the Anunnaki-gods wept with her; the gods sat bowed and weeping.") After The 12th Planet, Sitchin wrote many other successful books with such themes, including The Anunnaki Chronicles, The Cosmic Code, Genesis Revisited, and The Earth Chronicles. I wouldn't recommend taking any of them seriously.

As described at www.sitchin.com: "One of the few scholars able to read and interpret ancient Sumerian and Akkadian clay tablets, Zecharia Sitchin (1920-2010) based his bestselling The 12th Planet on texts from the ancient civilizations of the Near East. Drawing both widespread interest and criticism, his controversial theories on the Anunnaki origins of humanity have been translated into more than 20 languages and featured on radio and television programs around the world."

But at http://www.skepdic.com/sitchin.html we are told:

"Like von Däniken and Velikovsky, Sitchin weaves a compelling and entertaining story out of facts, misrepresentations, fictions, speculations, misquotes, and mistranslations. Each begins with their beliefs about ancient visitors from other worlds and then proceeds to fit facts and fictions to their basic hypotheses. Each is a master at ignoring inconvenient facts, making mysteries where there were none before, and offering their alien hypotheses to solve the mysteries. Their works are very attractive to those who love a good mystery and are ignorant of the nature and limits of scientific knowledge. They are especially attractive to those who are ignorant of biblical and historical scholarship.

Sitchin promotes himself as a Biblical scholar and master of ancient languages, but his real mastery was in making up his own translations of Biblical texts to support his readings of Sumerian and Akkadian writings."

The idea that there is a dangerous rogue planet called Nibiru which could swing by the Earth and cause great destruction on multiple occasions probably would not exist without the fiction written by Zecharia Sitchin. The idea is popular. Millions of people, especially those with little scientific background, are obsessed with this idea of "Nibiru." But this has almost nothing to do with what Nibiru really is, or what Nibiru meant to the ancient Sumerians. Perhaps this idea of Nibiru is summed up best by a 2015 article by Robert Walker on science20.com which he titled "'Imaginary Bullshit Planet' Nibiru – Lens Flares, Sun Mirages, Hoaxes, & Just Plain Silly."

Almost as if the world needed someone to make Zecharia Sitchin look like a credible scholar by comparison, in 1995, an alleged psychic named Nancy Leider claimed that, through an implant in her brain, extraterrestrial aliens from Zeta Reticuli (Zetas) were sending her messages about a rogue planet that would cause mass destruction when it would pass near Earth in 2003. She called it Planet X, and eventually said that it was the same planet known as Nibiru. Despite nothing happening in 2003, or 2012, or any other time frame "the Zetas" have claimed would be the time of a pole shift caused by Nibiru, she still has a cult following her that pays attention to the alleged instructions from the Zetas.

At www.zetatalk.com: we are told that "ZetaTalk leads you through the vast amount of information being relayed by the Zetas in answer to questions posed to their emissary, Nancy Lieder.

ZetaTalk answers cover such subjects as portents of a Pole Shift and how this relates to the Transformation in process; how life in the Aftertime following this shift will be different from today; the self-centered or service-minded spiritual Orientation of humans as well as aliens from other worlds and how inadvertently giving the Call to aliens can put you in touch with one group or the other; how Visitations can be more easily interpreted when spiritual orientation is understood; how visitors from other Worlds are watched by the Council of Worlds, which has set Rules regulating their behavior; why we are only gradually getting acquainted with our visitors from other worlds, and what will allow the Awakening to occur faster; to what extent the Government is aware of and interacting with the alien presence; the true nature and reason for the Hybrids being developed by the Zetas to merge the best from both Zetans and Humans; why aliens can disappear and move through walls, and what both physical and spiritual Density changes will be like in the future; what the Zetas have to say about our Science theories; what the Zetas as students of human nature have concluded on what Being Human means; and straight ZetaTalk about our Myths."

As Ken Adachi wrote online: "I had expected Nancy Leider and Zeta Talk to disappear from the Internet following the 2003 fiasco, but I guess a built up CIA/NASA/military disinformation psy-ops like Zeta Talk is simply too juicy a scam to just abandon, so they proffer a truly moronic and pathetic 'White Lie' excuse to explain away their prediction failure and just move the date up another [nine] years to 2012! So we are treated with another [nine] years of disinformation nonsense, fear mongering, anxiety promotion... for a whole new generation of suckers, rubes, and newbies who will buy into the Zeta Talk lie with the same wide eyed belief that I bought into it between 1995 and 2003, and wind up posting Youtube videos with ingenious sounding explanations of mysterious celestial body movements..."[255]

Perhaps one of those he describes runs poleshift.ning.com, a site devoted to the pole shift idea, but as I see it, heavily contaminated with a focus on Planet X and Zeta Talk.

More recently a "Christian conspiracy theorist" named David Meade has experienced more than fifteen minutes of fame for what seems like baseless nonsense to me, warning of impending doom from Planet X or Nibiru either gravitationally shifting the Earth's axis, or crashing directly into us. It was supposed to happen September 23, 2017. When that didn't happen, we must have misunderstood what he said. Now he says we should watch for various dates in October, all of which, I predict, will have passed by uneventfully long before you read this.

I understand that to some readers, I too am a conspiracy theorist writing nonsense about doomsday scenarios. It may provide entertainment to some skeptics as one prognosticator mocks another, when (hopefully!) the dates that I find potentially significant also pass by uneventfully. In my defense, I believe that many sciences show evidence of multiple previous pole shifts, and even Albert Einstein agreed with that much.

As for interpreting prophecy and attempting to use forensic astronomy and other analysis to calculate timelines and dates – I could be wrong, but I'm giving it my best effort. To ignore prophecy, especially with about a third of the Bible being prophecy, is tantamount to saying that the Bible is nonsense. It doesn't seem logical to me to ignore the idea that the generation that saw Israel recreated in 1948 would see the end times unfold – and to do so right as the years of that generation seem to be coming to an end. So when I think I have discovered patterns and meaning scattered through the early texts of many ancient cultures, I assume it is more meaningful than finding patterns in the random ink blots of a Rorschach test. Of course, the near future will be far more stable if I am completely wrong on timing.

But in this article from 2015, I explain why I think the people who expect a rogue planet to destroy the Earth are the ones who are wrong. From "Nibiru: It's Not What Most People Think It Is"[256]

"Nibiru – as understood by most modern readers – is complete B.S. And what I mean by that is that the idea of some exoplanet from beyond our solar system orbiting in every 3600 years and coming close enough to cause catastrophes on earth – yet magically not quite close enough to mess up Earth's orbit and kill us all in any of the last however many thousands of approaches – is not scientifically plausible.

Some of the best evidence against a large astronomical body ever having come close enough to Earth to cause any danger from its gravitational effects is the stability of our moon. The moon has a very stable orbit around the Earth, it is tidally locked such that its spin matches its orbit and one side continually faces our planet. It has had the same orbit and cycle for well over 3600 years – with many ancient lunar calendars documenting the details not just four or five thousand years ago, but all the way back to very basic Aurignacian lunar calendars estimated to be 34,000 years old.

The idea of regular and catastrophic near misses – in which a rogue planet repeatedly almost hits the Earth every 3600 years – was made up by the popular author Zechariah Sitchin, but the idea is nonsense.

There is a wonderful scholarly analysis of Nibiru by a PhD [Dr. Michael Heiser] here: http://www.sitchiniswrong.com/nibiru.pdf

Let's go straight to the source for some real insights on Nibiru:

In the Sumerian creation epic, the Enuma Elish, Nibiru is mentioned many times. The root of the word means "crossing point" and although it is usually used in conjunction with a god, a star, or another astronomical body, it is sometimes used to merely represent a bridge over a river, a gateway, or a door.

Nibiru is often referred to as the great Babylonian god Marduk's star or planet. It has been associated with Jupiter (many times) Mercury (once) the pole star, and an otherwise unnamed star. It is described as appearing irregularly and appearing every year, and is described as having a fixed position and as having changed its position. I believe this confusion is because the true importance of Nibiru is in the form of an alignment of multiple heavenly bodies. I believe this was understood by earlier Sumerian writers, but that the concept was no longer fully understood by later Babylonians who still used its name with single heavenly bodies.

If you have read my books you know I believe that one of prophecy's most important astronomical alignments will occur in the sky in late December 2019, when the winter solstice sun is in alignment with the galactic center – and is met by Jupiter. I believe the sun, moon, and planets move into positions which act out all the major steps of an ancient Jewish wedding ceremony – and that this has been described in the prophecies of many "myths" and religions, most importantly as Jesus returning for His bride…

This alignment by the galactic center is an astronomical marker in time – a clue to the timing of one of Dr. Paul LaViolette's galactic superwaves arriving based on the periodic activity (Seyfert phase) of the massive black hole at the galactic center. Around 1940, Carl Seyfert observed that roughly one in seven spiral galaxies (like our own Milky Way) have an active black hole at their core spewing out dust and light and other radiation like a quasar. "Astronomers have come to realize that galactic core explosions occur in all spiral galaxies, even our own, and that the majority of galaxies that have a normal appearance with no sign of core activity are simply galaxies whose cores happen to be in their quiescent phase…. Outbursts recur about every ten thousand years" or "perhaps every ten to twenty thousand years."[257] Each event ejects an expanding spherical shell which LaViolette calls a "galactic superwave."

As Earth is roughly 23,000 light years from the galactic core, it is very likely there are at least two such superwaves expanding out towards us right now – but because they move at light speed, we are beyond their event horizon; we cannot see them coming.[258] We can see them in other nearby galaxies like Andromeda (at least 3 superwaves can be detected expanding out from its center) and Centaurus A (with as many as 18 superwaves); and we can see evidence that they have passed by us before – for example, we can analyze supernova to find many that were trigger by the same passing superwave, and observe that their debris is not spherical but distorted as if blasted from the direction of the galactic center. But we cannot detect the next incoming superwave without some kind of "warp" technology that would allow us to go faster than light. Ancient records may warn us of the timing, but without accurate ancient knowledge based on past cycles we will only know when we see it and feel the next superwave arrive.

This is the astronomical representation of God becoming manifest in the sky, because just as the cosmic burst of radiation (including visible light and gravity waves) arrives and triggers a pole shift, the galactic center suddenly becomes visible, even in daytime. The survivors of the pole shift, and many generations of their descendants, will see a bright blue "Eye of God" in the sky, with a galactic center "pupil" and a dimmer blue oval around it, as light is reflected off dust in the central bulge of the galactic disk for a period which Dr. Violette estimates at about 1,000 years. The longer path of reflected light would take extra time to reach us as some blue light bounces off dust clouds near the galactic center, and near our own solar system, illuminating various dust clouds and reaching us over the course of the centuries following the initial superwave arrival and pole shift. Dust clouds near our solar system would appear like shimmering, ghostly blue auroras all over the sky, while the density of clouds near the galactic center would accentuate the giant "eye in the sky." Dr. LaViolette estimates that the galactic center "pupil" would dominate a dimmer "iris" about eight times the apparent width of the sun or moon... which may have "appeared to ancient inhabitants as a gigantic punishing 'Eye' in the sky, the entire form occupying about a 16-degree field of view, or about 32 solar diameters."[259]

Was the eye considered the visible presence of God? The verses of Genesis 3:7-10 are interesting in this context. This is just after Adam and Eve have eaten from the tree of knowledge, and "the eyes of both of them were opened." Is it relevant that eyes are associated with this new understanding? Verse 3 immediately adds that "They heard the sound of the Lord God walking in the garden in the cool of the day, and the man and his wife hid themselves from the presence of the Lord in the garden." Verse 10 makes it even more clear when Adam says "I heard the sound of You in the garden, and I was afraid." God's presence was audibly and visibly present at this time, and with their sudden new understanding they were for the first time scared of God's presence. By the end of Genesis 3 they are driven out of Eden: "So He drove the man out, and at the east of the garden of Eden He stationed the cherubim and the flaming sword which turned every direction to guard the way." If this is a distorted tale of surviving a pole shift, it makes some sense.

Before the pole shift, life was good and stable, in a nice climate with plentiful food. There may even have been the comforts of an advanced civilization. At the moment man was experiencing a great new understanding of things, God appeared and was visibly and audibly present, and man was terrified. In a catastrophic pole shift with violent earthquakes, tsunamis, and rapidly changing skies, most of the world suffers a major change in climate. Most of those who survived the initial changes in latitude, altitude, and temperature may have felt the need to evacuate their homeland before they would be overwhelmed by rising waters, falling snow, or other new pole-shift-related problems. A bright blue eye and ghostly auroras in the sky would be very real reminders of God's recent activity.

Other references make it clear that God was once present with man. The Lord appeared to Abraham and Moses, and this is referenced in the phrase: "May the Lord our God be with us, as he was with our fathers." King Solomon, around 3,000 years ago, lamented the absence of God when he said this in 1 Kings 8:57, as he was about to consecrate the new Temple in Jerusalem with a week-long feast.

I believe this corresponds to the creation of the New Jerusalem, and the new heaven and new earth described in the Book of Revelation, which exist after a pole shift. The measurements of the Book of Revelation's New Jerusalem, using "human measurements, which are also angelic measurements" (Revelation 21:17) are very insightful. There are important ratios between the two, disguised from us in unfamiliar units of cubits and furlongs... as opposed to "the so-called English foot" which is really a very, very ancient unit of measure. "Dividing the equator into 360 degrees makes each degree equal to 365,243.22 feet, or the number of days in a thousand years."[260] I can't help but think of 2 Peter 3:8-9 "do not let this one fact escape your notice, beloved, that with the Lord one day is like a thousand years." (And I can't help noticing that "English" units of measure are ancient units used in pyramids and monuments around the world, and that the units themselves were based on fractions of the known sizes of the Earth, moon, and sun, and the distances between them.)

John's "New Jerusalem is not of human manufacture, and the units of measure that pertain to its dimensions are properly called sacred because they derive from eternal standards in number."[261] When you realize that 14,400 cubits are 24,883.2 feet and that the circumference of the Earth is 24,883.2 miles... or that 12 furlongs are 7,920 feet and the diameter of the Earth is 7,920 miles – you begin to see that the measurements of the New Jerusalem may just be a thinly veiled reference to measurements of the Earth, and part of a description of a perfected and newly re-created Earth, to experience another golden age after the next pole shift.

And what is veiled as distance to measure may very well involve a measurement of time – the time between pole shifts. If you realize that there are 5,280 English feet in an English mile, but that a Greek mile was only 5,068.8 English feet, you might do a little math and realize that the circumference of the Earth is 25,920 Greek miles[262] – the same number as there are years in a cycle of precession – and you might assume that there are clues to timing such pole shift events as well, all hidden in the earthly and angelic measurements of the New Jerusalem.

I suspect the creation week in Genesis also describes a pole shift, and that the week-long wedding ceremony that will be acted out in the sky in December 2019 over the week-long holiday of Hanukkah (celebrating dedication of the temple) may all relate to the same type of pole shift events. I am not saying I am certain that a pole shift is due on any particular date – there are also some clues pointing to later dates... but I am watching that time frame as one of the most likely times, indicated by many prophetic clues... Regardless of the exact timing – there are probably two galactic superwaves in our galaxy expanding outwards towards us right now, as there always are. It is just a matter of when the next one arrives. Did our ancient ancestors encode their clues and warnings in a way that successfully allows someone like me to calculate the arrival of the next superwave event, using descriptions of astronomical alignments as markers in time?

Dr. LaViolette says that mythology and "astrology's symbols convey a highly sophisticated astronomical and geological message, one that informs us and future generations about one of the most horrible disasters to afflict the human race – the occurrence of an explosion of our Galaxy's core. Moreover, it warns us that this tragedy could repeat."[263] I do believe that our ancestors tried to make this wisdom available to future generations, if only to the initiated... Randall Carlson (and many others) says there have been "esoteric orders, or brotherhoods, or mystical organizations, however you want to think of them, whose main purpose has been to preserve and transmit these traditions of lost knowledge... the knowledge of the great cycles and the great events that have caused the termination of these cultures." There could be a brotherhood still carrying ancient messages forward today, who know the ancient wisdom for certain – whereas I am merely speculating that the Seyfert phase of spiral galaxies is the trigger of our pole shifts, and that the blue "eye in the sky" will only be witnessed after the cataclysm...

The last time the active phase of our galactic center was visible, I believe it was known to the Egyptians as the Eye of Horus; in India as the Navel of Vishnu; to Hebrews as the Shekinah Glory; to Muslims it is remembered as the Tariq Star; to the Hopi it is the Blue Star Kachina... And I believe that to the Sumerians, it was remembered as "Nibiru."

It sits at the crossing point of the Milky Way (the "spine" of the galaxy provides a visible line in the sky) and the path of the ecliptic – the course/line followed by the sun and planets through the signs of the zodiac. At this crossing point of the two lines is the galactic center, which periodically becomes brightly visible at the time when an incoming superwave arrives and causes destruction like a pole shift. This is the Mayan conjunction of the four roads in the sky and when the winter solstice sun (the sun on its shortest, weakest, dying day) appears at this crossroads it is represented by Jesus dying on the cross and being reborn a few days later.

So when the Enuma Elish cuneiform tells us "Nibiru is his [Marduk's in context] star, which he made appear in the heavens" (Tablet VII, line 130-131) we know Nibiru "appears" – it had not been visible, then it becomes visible, like the galactic center during a Seyfert phase...

When the Enuma Elish says "let Nibiru be the holder of the crossing place of the heaven and of the earth" (Tablet VII, line 124) we know Nibiru sits at an important crossing place in the sky – in my opinion, the same crossing point of the four roads in Mayan mythology, as "their crossroads format suggests the astronomical location of the crossing point of the Milky Way and the ecliptic."[264] When the Enuma Elish tells us Nibiru is "the seizer of the midst" and is "He who forced his way through the midst of Tiamat" we should recognize that the Sumerians' cosmic serpent or Leviathan, the goddess of chaos – Tiamat – is astronomically represented as the Milky Way, much like the sea of glass with four living creatures (zodiac signs) in the midst of the throne in Revelation 4:6. Nibiru is described as seizing the midst of the Milky Way because Nibiru suddenly appears in and illuminates the central bulge of the Milky Way.

In the Bible, Jeremiah 6:1 mentions "the midst" and "a sign of fire" and "great destruction." It mentions the central location of Jerusalem, which may be represented astronomically at the galactic center. In verse 16 God asks us to "stop at the crossroads and look." Are we supposed to look at the crossroads in the sky, the spot Sumerians called Nibiru, where the ecliptic and Milky Way cross paths, for the sign of fire and great destruction? Verse 16 ends with people saying they will not look, and in verse 17 God says: "Then I will appoint watchmen to direct you." Are books like this one, or ancient prophecies, or initiated brotherhoods here to direct your attention to the galactic center?

As the galactic center is normally not visible, but is brightly illuminated at the start of horrific events that periodically cause a pole shift – which happens to occur when the winter solstice sun is near the galactic center in conjunction with Jupiter – this conjunction is also associated with our sun, and with Jupiter, and with the pole shift that sets a new course for the heavenly bodies in the sky. The heavenly bodies do not actually change course, but because the ground beneath us moves in relation to our axis of rotation the course of heavenly bodies appears to change, from the view of survivors of the pole shift.

"He [Marduk in context] set fast the position of Nibiru to fix their [the stars] bounds" (Tablet V, Line 6) – the galactic center is the central and most important point in the sky, newly visible and appearing at the time of destruction of the former world, so the crossing point it marks must seem very important. This new eye of God appearing in the sky must seem to be responsible for setting the new course of the stars and planets in the sky.

In Arthurian legend, the rightful king attains his kingship when he can pull the sword (Excalibur) from the stone. (This also refers to pulling hidden knowledge – ancient, esoteric, astronomical wisdom – out of the basics of Christianity, the rock the Church was based on. See page 336 of <u>End Times and 2019</u> for the Kabbalistic view of this.)

Does God grant kingship over the earth when the sword/axis is removed from the stone/planet? Much like the Book of Revelation describes Jesus gradually approaching and eventually reaching the throne of God, to sit as his right hand and receive kingship over the Earth, (which I believe also describes the slow precession of the winter solstice sun towards the position of the galactic center) – or the Mayan return of their sun-god Kukulkan to receive kingship over the Earth when the Pleiades and the sun come together in the zenith directly over Chichen Itza (which happens every year for several centuries out of the 25,800 year precession cycle – starting in the 21st century) – likewise "Marduk is singled out as the champion of the divine court, undertakes the commission and is invested with kingship to sustain the royal authority of Anu and the heavenly court against the menace of primeval Chaos..." After defeating the forces of Chaos, led by Tiamat, Marduk then reestablishes the proper movement of the stars "After defining the days of the year by means of heavenly figures, He founded the station of Nibiru to determine their heavenly bands, That none might transgress or fall short."[265]

One other mention from outside the Enuma Elish is also worth noting, as it describes the central position (in the central bulge of the Milky Way) and controlling role Nibiru plays in the new order in the sky when we are told "after the gods of the night [the stars] have been finished, dividing the sky in half, this star is Nibiru."[266]

In summary, I believe Nibiru originally represented a god-star-crossing-point – the crucial alignment which occurs with Jupiter and the sun at the suddenly visible galactic center during a pole shift, but that this understanding was lost already by Babylonian times, because the galactic center only remains visible for a few centuries. Eventually the correct understanding of Nibiru, and how the galactic center's periodic active phases lead to pole shifts, was lost. The correct understanding was lost gradually over time – leading to improper use of the name Nibiru for multiple heavenly bodies, causing much confusion, greatly compounded by the false teachings of Sitchin in modern times.

Not every dot of light in the night sky means Nibiru is approaching. Not every lens flare in a photograph is Nibiru. Statistically, any rogue star or planet crossing near enough to Earth to cause devastation every 3600 years (for which there is no serious evidence) would have entirely destroyed the planet ages ago. Astronomers in China and Egypt and other advanced, sky-watching cultures would have recorded the event, not just the Sumerians with only Sitchin being capable of interpreting their story… While I do think the real "Nibiru" (the galactic center) will be visible in the sky soon because I think evidence points to a galactic superwave and pole shift coming soon, I merely believe Nibiru refers to the newly visible galactic center experiencing its active/Seyfert phase, not some non-existent rogue planet allegedly approaching us.

If you are willing to study the past in order to understand the great cycles that are about to impact our very near future, consider reading – Earth Under Fire – by Dr. Paul LaViolette, Maps of the Ancient Sea Kings by Professor Charles Hapgood, The Dimensions of Paradise by John Michell, and The Great Pyramid Decoded by Peter Lemesurier. For deeper insight into the specific astronomical alignments described in the Bible (and other ancient sources) that occur in December 2019, consider reading End Times and 2019."

I soon posted another article on the same web site (Beforeitsnews.com): "NIBIRU – It's NOT a Rogue Planet bringing DOOM to Earth"

"At least, that's not what the Sumerians meant when they first wrote about Nibiru.

Readers do seem fascinated with the idea of NIBIRU. I explained that it is NOT what most people think (a rogue planet in an unusually long orbit around our sun) in an article a few days ago that has had many readers, including many with comments defending the silly pseudo-science I debunk.

Just to be clear, I am not stating that there are no such planets in a long orbit around our sun – "Planet X" could be discovered in orbit beyond Neptune and Pluto tomorrow. I am merely saying that such a planet does not come close to Earth and cause catastrophes every 3,600 years, and that no planet orbiting beyond Neptune would be the home world of alien gods either.

John Ale has been posting several Nibiru related articles on BIN in recent days including "Planet X – Nibiru Validated, Sitchin Vindicated – Scientific Evidence." REALLY?

Ale says: "'Planet X' has become the term associated with the elusive planet(s) and as a result has become a synonym for what Ancient Astronaut Theorists refer to as Nibiru."

NO IT HAS NOT. That is making a huge jump from stating that more "trans-Neptunian" objects (like Pluto and Sedna and the Oort Cloud and the Kupier Belt) probably exist to be discovered soon – to being "a synonym" for Nibiru and Ancient Astronauts.

About six minutes into Ale's video ("Planet X Science – Nibiru Validated, Sitchin Vindicated)[267] Lord Pye is quoted describing an alleged near-collision between Nibiru and Saturn (Tiamat) which ended in Nibiru stealing away one of Saturn's moon's before releasing it (Pluto) in its own orbit. This is presented as "fact." Later the video describes Pluto as a lost moon of Neptune. Then around the thirteen minute mark the video says "the discrepancy does not take away from the big picture. In essence, on the visit to the inner solar system in which Nibiru collided with Tiamat (now Earth) it was knocked off its elliptical orbit." What!?!? Discrepancies in the story don't matter, and now Tiamat isn't Saturn but Earth?

But since Ale's recent articles clearly focus on Nibiru as described by Zechariah Sitchin and his kooky theories while claiming such nonsense is "vindicated" – let's examine what Sitchin actually claims:

...Wikipedia's entry for him begins: "Zecharia Sitchin (Russian: Захáрия Ситчин) (July 11, 1920 – October 9, 2010) was a Soviet-born American author of books proposing an explanation for human origins involving ancient astronauts. Sitchin attributes the creation of the ancient Sumerian culture to the Anunnaki, which he states was a race of extraterrestrials from a planet beyond Neptune called Nibiru. He believed this hypothetical planet of Nibiru to be in an elongated, elliptical orbit in the Earth's own Solar System, asserting that Sumerian mythology reflects this view. Sitchin's books have sold millions of copies worldwide and have been translated into more than 25 languages. Sitchin's ideas have been rejected by scientists and academics, who dismiss his work as pseudoscience and pseudohistory. His work has been criticized for flawed methodology and mistranslations of ancient texts as well as for incorrect astronomical and scientific claims."

As a citation we are sent to another site[268] which tells us "Sitchin's claim to fame is announcing that he alone correctly reads ancient Sumerian clay tablets. [Of course, he didn't announce this by taking out an ad in the New York Times but by implying it with his "translations" that do not jibe with the work of legitimate scholars in the field.] If Sitchin is right, then all other scholars have misread these tablets, which, according to Sitchin, reveal that gods from another planet (Nibiru or Niburu, which orbits our Sun every 3,600 years) arrived on Earth some 450,000 years ago and created humans by genetic engineering of female apes. Niburu orbits beyond Pluto and is heated from within by radioactive decay, according to Sitchin."

So let's get this straight – Sitchin says alien gods from a distant planet in our solar system genetically engineered female apes long ago and founded the human race. He says their home planet is heated by internal radiation (Does anyone comprehend how much radioactive decay would have to happen to heat a planet beyond Neptune several hundred degrees warmer? And how lethal that would be to life forms? And that even if it were warm on such a planet, there would be almost zero light for photosynthesis? Meaning no plants to grow, which means no food for animal life like us or like the alleged radiation-proof alien gods?)

Dr. Michael Heisner wrote a long article on what the Sumerians really meant when they were talking about Nibiru. It can be found at http://www.sitchiniswrong.com/nibiru.pdf and the link/title clearly shows that he feels Sitchin is wrong.

Quotes from that document include:

The Myth of a Sumerian 12th Planet: "Nibiru" According to the Cuneiform Sources

Michael S. Heiser, Ph.D. Hebrew Bible and Ancient Semitic Languages University of Wisconsin-Madison

it begins:

"Those familiar with either the writings of Zecharia Sitchin or the current internet rantings about "the return of Planet X" are likely familiar with the word "nibiru". According to self-proclaimed ancient languages scholar Zecharia Sitchin, the Sumerians knew of an extra planet beyond Pluto. This extra planet was called Nibiru. Sitchin goes on to claim that Nibiru passes through our solar system every 3600 years. Some believers in Sitchin's theory contend that Nibiru will return soon – May of 2003 to be exact. These followers of Sitchin's ideas also refer to Nibiru as "Planet X", the name given to a planet that is allegedly located within our solar system but beyond Pluto. Adherents to the "returning Planet X hypothesis" believe the return of this wandering planet will bring cataclysmic consequences to earth.

Is Sitchin correct – Is Nibiru a 12th planet that passes through our solar system every 3600 years? Did the Sumerians know this?

Are those who equate Sitchin's Nibiru with Planet X correct in this view? Unfortunately for Sitchin and his followers, the answer to each of these questions is no."

Sitchin of course notes the basic "crossing" meaning in his book. One just needs a dictionary for this, as the above indicates. He then supplies – without textual support – the idea that Nibiru is a planet that "crossed" paths with other planets in our solar system on its regular 3600 year course.

...Nibiru-Marduk-Jupiter has something to do with a crossing place – that is, this astronomical body itself isn't DOING the crossing, but marks or is positioned at a crossing point. This is another point of contradiction with Sitchin's teachings, as he argues it is Nibiru that is mobile and "crosses" into the orbital paths of our solar system's planets. The texts do not say this. [Ancient descriptions of Nibiru are more in line with the Mayan view: "the sacred ruler does not just sit on the throne symbolizing the cosmic center and crossroads; instead, he is the crossroads"[269] which are also symbolic of a black hole, of kingship, and of rebirth.]

Tablet VII then continues in lines 126-127 with this statement: "Nibiru is his [Marduk's] star, which he made appear in the heavens. He [Nibiru] is the one who holds the "turning point" [the crossing point, juncture]; they must look to him."

Observation: Horowitz notes that the word kunsaggu is only attested in Enuma Elish. There is a kun-sag-ga in another text where that word is paralleled by the word muhru, which is a street sign or marker for a turning point during a processional circuit. At any rate, Nibiru marks some sort of juncture or intersection."

The final reference to Nibiru in Tablet VII is lines 130-131, where the name does not occur, but which continues the thought of lines 126-127: The stars of heaven, let him [Nibiru] set their course; let him shepherd all the gods like sheep."

Observation: This "station" is explained by scholars by referring to the above lines (124-127) in Tablet VII. Nibiru's "stationing" is taken to be the "special role" assigned by Marduk: some sort of regulating point or influence over the stars that seems to dictate the courses of the stars."

Nibiru is never mentioned in any respect with the Anunnaki; it is never said to have been or be inhabited.

...Nibiru was [long ago] seen every year, which demolishes Sitchin's view of a 3600 year cycle for it."[270]

I think the most likely explanation for Nibiru is that it referred to an astronomical conjunction at a crossing point. I suspect the crossing point is at the galactic center, where the ecliptic crosses the spine of the Milky Way. I suspect the conjunction involves the winter solstice sun, and Jupiter, at the galactic center when it becomes visible again as a bright blue "star" during a

Seyfert phase or galactic superwave event. This would correspond to the Maya concept of the throne manifesting into existence only when the December solstice sun joins with the galactic center, when their sun-god Kukulkan or One Hunahpu sacrifices himself for humanity's renewal at the center of the four crossroads, or Jesus sacrificing Himself on the cross.

By Babylonian times, it was still understood that global cataclysms occur when Nibiru appears, but what Nibiru was exactly seems to have been lost in translation and time, eventually too closely linked to Jupiter, Marduk, and even Nergal. Nergal, (sometimes known as Erra) the Babylonian god of death, war, destruction, and mayhem usually depicted as a lion – was once summoned by the seven Sebetti (the sons of heaven and earth – much like the Greek Titans) to destroy mankind. Are these the stars of the Big Dipper – the Seven Rishis of India who beget twins (new poles?) on the Earth?

And Nergal/Erra commented on eventually being ready to end the world again and said of the next cataclysm: "Open the way, I will take the road, the days are ended, the fixed time has passed." This seems to indicate that Nergal's ability to end the world will only be possible at the end of a certain time period or cycle, when "he" has moved along a "road" and reached a certain destination. Nergal sounds like an astronomical body, and several sources say he was the summer solstice sun or even "the summer solstice that brings destruction." (see Wikipedia on Nergal) Half a precession cycle back, almost 13,000 years ago at the beginning of the Age of Leo, the summer solstice sun was moving along the path of the ecliptic in the cycle of precession to reach the galactic center at the time of the last pole shift…. Nibiru was probably the newly active, lit up galactic center at the time of the pole shift when Nergal was in conjunction at the crossroads of the ecliptic and the Milky Way…

After Nergal's comments on timing, the Babylonian high god Marduk (often associated with Jupiter, like Zeus and many other top gods) then says "When I stood up from my seat and let the flood break in, then the judgment of Earth and Heaven went out of joint… the gods, which trembled, the stars of heaven – their positions changed, and I did not bring them back." Marduk sounds empowered by Nergal to cause the destruction. Marduk "stood up" – he moved and made himself obviously seen – and then let everything fall out of place and did not bring it back – Marduk caused a new path for the stars.[271] I believe that Marduk, as Jupiter, may have been a third part of a conjunction between the galactic center, summer solstice sun, and Jupiter comprising "Nibiru" at the crossroads.

Nibiru was identified as the summer solstice by Ernst Weidner in his 1915 book: Handbuch der Babylonischen Astronomie on page 33... F.M.T. Bohl wrote an article in 1936 on the "Names of Marduk" in which he said that Nibiru was a "hypsoma" (hidden place) and the location of an alignment "marking the point where Jupiter entered 'the Way of Anu'" (The Milky Way) …"when the Sun stands at the crosspoint of Equator and Ecliptic." The "hidden place" reference makes sense to me, as the galactic center is hidden from view except during its active phase for up to about a thousand years after the arrival of a galactic superwave. In Heidel's translation of the Enuma Elish, "Nibiru is the star which they caused to shine in the sky. He has taken position at the solstitial point." Stephen Langdon's 1923 book: The Babylonian Epic of Creation suggests on page 156 that "It is on the whole clear that Nibiru (the crossing) refers to the intersection of the celestial equator and the ecliptic and that the name was applied to Jupiter as a representative of the planets which cross from the southern to the northern part of the Way of Anu."

Despite my cherry-picking supportive quotes to back my own theory – that Nibiru originally meant the newly visible galactic center at the crossroads of the ecliptic and the Milky Way, probably in alignment with the summer solstice sun and even Jupiter – there is plenty of evidence supporting my theory. But overall there is no scholarly consensus on what Nibiru initially represented; there are simply too many changes from early Sumerian to later Babylonian references.

As one modern article on the subject concludes: "the detailed assignment of visible markers of the Crossing to astronomical objects may have evolved over time."[272] As I explain it: by Babylonian times they no longer fully understood the original references to the astronomical alignment that was Nibiru, which were already ancient references subject to mistranslation in Sumerian times.

My best guess that Nibiru was an alignment of the summer solstice sun and Jupiter with the galactic center at the heavenly crossroads of the ecliptic and

Milky Way around 12,900 years ago – is merely my best guess... But it corresponds amazingly well with the astronomy half a precession cycle later on the date I was led to while writing another book. I had no interest in Nibiru back then; I don't think the book even mentions it – but I am reminded of the astronomical lineup emphasized in End Times and 2019 on a potential Judgment Day in our near future – when the galactic center, the winter solstice sun, and Jupiter are in alignment.

Many pole shift scholars suggest the next catastrophe is due "soon" but soon might not even be this century for some of them. Nostradamus suggests we should expect a pole shift around 2028-2029. I hope my timetables are way off in both Nostradamus and the Islamic Invasion of Europe and especially in End Times and 2019. If we are lucky, the next pole shift will not occur in our lifetime. There is a lot of evidence for periodic pole shifts occurring in cycles of approximately 12,900-12,960 years... and the last one was approximately that far back in the past. But the margin of error is large; and whether we are due in one year or a hundred years or more, I cannot say with certainty. I can only assess the evidence from many sources, including the clues left to us by Mystery Schools and religions, and speculate on possible interpretations and timing. But if we are going to watch for a pole shift involving Nibiru in the near future, I think we should watch for the upcoming astronomical alignments of late December 2019 – when the galactic center, the solstice sun, and Jupiter are not only aligned in the right place, but their relative movements help act out a heavenly wedding ceremony.

World Mythology

Chinese Mythology

In Chinese mythology, we are told that "at the beginning of the second heaven, the earth was shaken to its foundation. The sky sank lower and lower towards the north. The sun, moon, and stars changed their motion. The earth fell to pieces and the waters in its bosom uprushed with violence and overflowed it." This story of the "second heaven" is probably not referring to the northward displacement of Asia during the last pole shift, because the sky would have appeared to sink into the south while Asia moved north. The description may refer to the pole shift before that, when the North Pole moved from the Greenland Sea to Hudson Bay.

Other Chinese myths tell us the last world ended when an evil monster tore down the heavenly bamboo (world axis) by "tilting the earth," flattening mountains and tearing a hole in the sky. The waters flooded the world, fires raged across the land, and the cardinal directions of north, south, east, and west fell out of alignment. "Earth failed to hold up heaven all the way around [its circumference.] Why were the eight pillars too short for it in the southeast?" (This gives valuable insight into WHERE the ancient ancestors of modern China felt "the problem" that led to the sky falling was caused or the direction of the falling portion of the stars. Viewed from China – and anywhere else north of the tropic of Cancer – the sun is always supposed to rise in the southeast. It didn't.

And if the North Pole was in or just west of Hudson Bay, and the people creating this story were in what we now call China, the "problem" that caused the pole shift lay in the general direction of the ice sheet in East Antarctica – close to where the same problem lies today. This was southeast of China, before Asia shifted north in the last pole shift. After the sky fell, Chinese myths tell us that Nuwa (or Nu Kua) fixed the pillars of heaven, fixed the cardinal directions, established a new order and repaired the sky. Fu Hsi (or Fuxi) then measured the roundness of heaven and the squareness (based on four cardinal points) of the earth. Pillars were set up and waters drained off the land. In the Pangu (or Pan Ku) version of the Chinese creation myth, it took 18,000 years for the full heavenly cycle to complete. Considering how the facts can be corrupted over thousands of years, (such as estimating time periods based on generations) I think we should not ignore a number "in the ballpark" of the approximately 13,000 years of half a precession cycle that I suspect passes between pole shifts.

"The ancient Chinese encyclopedia Sing-li-ta-tsiuen-chou also refers to a Great Year, which it says measures the span of time between two successive catastrophes. It states that 'in a general convulsion of nature, the sea is carried out of its bed, mountains spring out of the ground, rivers change their course, human beings and everything are ruined."[273] Han dynasty astronomers attempted the noble work of calculating the exact length of these cosmic cycles, but they did so based on the somewhat arbitrary supposition of a recurring alignment of the new moon, winter solstice sun, and the five planets – only to come up with supercycles of no less than 138, 240 years.[274] Perhaps in an effort to stay in the ballpark of this figure, the 11th century Neo-Confucianist philosopher Shao-Yung taught that one cycle lasted 129,600 years. The Chinese believed ten such ages had passed from the beginning of the world until the days of Confucius.[275] One of the ten Chinese ages included a "sudden, flying leap to the Arctic."[276] Of course if we divide 129,600 years by ten ages instead of multiplying, we get exactly the 12,960 years in half a precession cycle that I assume is the true span of time between catastrophes. Was Shao-Yung trying to leave us a clue to this timing?

Greek Mythology

In Greek mythology, there is a great year of roughly 26,000 years (one precession cycle) with a great summer of destruction by fire (ekpyrauses) and a great winter of destruction by ice (kataklysmos) separated by about 13,000 years. Uranus was the original king of the gods in Greek myth. In the first epic battle between the gods, Uranus cast the Cyclopes and Hecatoncheires down to Tartarus much like stars falling below the horizon. The imprisonment is supposed to last many thousands of years, but is not meant to be permanent Gaia, mother of those cast down, eventually encourages Uranus' son, the Titan Chronos, to rebel against Uranus. Chronos' rebellion succeeds and liberates the banished gods from the previous war in heaven; though at some future point he casts them down again.

Tartarus is first visited by the mythic Greek hero Odysseus in his journey to Hades. Krates of Pergamon suggested some centuries later in his commentaries on mythology that the directions describing Circe's island and the journey to Hades and Tartarus really represent going from the Tropic of Capricorn to the South Pole. This makes sense of Dante's description of Odysseus' journey past the straits of Gibraltar, "always gaining to the left-hand side" – veering south along the west coast of Africa. Dante also describes the view from his ship: "all the stars of the other pole had come into sight, and those of ours had sunk so low that they did not rise from the sea."

Homer wrote the Odyssey centuries before Plato mentioned Atlantis. Yet Odysseus is shipwrecked on an island called Ogygia, home of the nymph Calypso. Calypso is described as the daughter of Atlas, who was the ruler of Atlantis. Ogygia was a king who ruled during a great flood. Plutarch wrote that beyond Ogygia is a great continent which surrounds the ocean. If you view the world from Antarctica on an azimuthal map centered on the South Pole, Antarctica seems to sit in the center of a world ocean, surrounded by a ring of northern land at the perimeter from Greenland and North America through Asia and Europe – matching Plutarch's description.

The Odyssey also brings the Greek God Hermes to Ogygia, who complains of its remoteness: "Zeus sent me here, I came against my will; who would choose to cross such an immense expanse of sea? There is not even a town of mortals around here." Odysseus himself says that when on this island he could "perceive neither where darkness is nor where dawn is, nor where the sun shining on men goes down underground nor where it rises." He said the rising sun "dances" which could describe its low movement along the horizon when viewed from an arctic or Antarctic position. So this land, associated with the great flood and with Atlantis, lies far across the true ocean and is uninhabited.

An Antarctic destination also makes sense of the more ancient version of the tale with silver pillars, (of ice) the cold waters of the Styx, and the disappearance of stars south below the horizon after a pole shift. Of the way that anything gets cast down there, Kleombrotos suggests conceiving of a river "whence flowed Time." Time itself arrives at this strange destination "once in ten thousand years." (Plutarch commented on this in <u>De defectu oraculorum</u> in 422 B.C.) This is a fairly close approximation to the 12,960 year half cycle of precession which I suspect is the time frame between pole shifts... The fact that the galactic center has a blue colored light in its active phase, and that Styx is associated with the sacred color of lapis lazuli in Greece, and with turquoise in Mexico, also suggests that the River Styx, associated with Death, may also be associated with the valley of the shadow of death between Sagittarius and Scorpio at the galactic center.

Chronos, the head god of the first Greek pantheon, was warned (in a prophecy) that his own son would eventually overthrow him, so he ate all his children, except for Zeus, who was saved without his knowledge. (In a different myth, Lykaon serves Zeus a dish of human meat from Zeus' own cooked son, enraging Zeus into "tilting the table" which caused the Flood of Deucalion – which sounds like a pole shift to me.) Zeus later led an alliance of Olympians against Chronos and the other Titans. I think it noteworthy that the Titans include Mylinos, "the Miller" and a leader of the Giants/Titans. Did Zeus upset the mill of rotation in this cosmic battle?

This war, known as the Titanomachy, was described in many classical Greek poems. The Olympians defeat Chronos, Zeus kills Mylinos and takes over the Mill of Heaven. Zeus also kills many other enemies, and forces the Titan named Atlas to bear the world on his shoulders. His territory of Atlantis, if in fact it is the same land as Antarctica, is the land depicted resting on Atlas' shoulders in artwork even today. Is Greek mythology describing repeated pole shifts and changes in the view of the heavens, when perhaps thirty degrees out of the sky's full field of view falls below the horizon, making it impossible to see certain stars again from the same location for thousands of years?

I find it interesting that in many cultures, the original pantheon of gods ruled on earth and dwelt among men, much as God originally walked among Adam and Eve in the Garden of Eden. Is this because for about a thousand years after the arrival of a galactic superwave, a giant "eye" is seen at the galactic center, and people feel God is visible and present and watching them? When that light in the heavens fades and "God" is no longer so visible, is he perceived to have left? Kronos ruled on earth, Zeus did not. Egypt's Ptah ruled on earth, Osiris did not. Or perhaps in the golden age, the earth's axis was viewed as right and correct in regard to the heavens, and after a pole shift it was viewed as warped or imperfect. Ptah is the only Egyptian god drawn with a straight royal beard; all others have a bent beard, and even this could refer to the bent or angled or shifted axis during the present, post pole-shift age.

Norse Mythology

Recent movies may have familiarized us with some characters of Viking mythology like Thor, but such movies hardly convey the battles of the Norse gods as descriptions of pole shift events. I contend that whenever a nation's mythology describes an epic battle amongst two sets of rival gods, we should consider that there really were dramatic changes in the heavens that led to such stories, with one set of heavenly bodies being "cast down" as our ancestor's position on the surface of the Earth suddenly changed in relation to the axis of rotation.

Norse mythology describes a previous battle between the Aesir, and the Vanir & Jotuns – but the future battle of Ragnarok is much more widely known. The gods will once again fight each other to the death, when the god's kingdom of Asgard is attacked by Frost Giants and other celestial forces from Muspell, a fiery domain responsible for creating the sun and stars (the galactic center?) and destroying the world. The very word Muspell may be derived from mund-spilli, world-destroyers. Ragnarok occurs when Muspell is flaming bright and the dragon (Draco) gnaws through the root (pole) of the world tree (the earth's axis?) Yggdrasil.

The Sun and Earth will barely survive the cosmic war: "The whole surface of the earth and mountains tremble so violently that trees are uprooted from the ground, mountains crash down… The Fenris-wolf then gets loose (to devour the sun)… Midgard, the serpent that encircles the world, now rears itself up from the sea… the seas wash over their banks, sweeping away all that remain of the world's inhabitants… (tsunamis) The Midgard serpent blows so much poison that the whole sky and sea are spattered with it… (cosmic dust) Surt flings fire over the whole earth and all things perish. (solar outburst) The

World Tree is wasted (axis shift) in the blaze... Stars fall, and darkness comes down upon the world. The seas flow over the burnt and wasted earth..."

Other related references in Norse myth include Odin's Eye, which was placed at Mimir's Well of Wisdom (the winter solstice sun at the galactic center?) and we are warned that this astronomical conjunction of the sun and a newly visible galactic center will happen again: "Odin's eye appearing as a visible object in the sky during times of calamity... the sudden appearance of Odin's eye was a presage to a world-encompassing disaster."[277]

Does the sun die for a bit before being reborn in Norse myth as well? We are told that the dragon Nidhug chews through the roots of the world tree (often symbolic of the axis, or the Milky Way) Yggdrasil, in which Odin (the sun?) once hung himself in sacrifice. Heimdall's horn is blown like a trumpet to announce the world's fate, and Jorgungandr – the Midgard Serpent – rises from the ocean depths and thrashes the waters violently in an attempt to find dry land, causing the seas to flood the world. Heimdall also has some interesting and little known nicknames like Vindler, which is derived from vinda – to turn, twist, or wind. If Heimdall does this he may represent the axis. He also holds the title of Hallinskidi, which is a bent stake or post – a clear reference to the axis bending.[278] The story has all the common elements of a "mythical" pole shift description so far...

Thor kills Jormungandr, but it splatters the sea and sky and Thor with so much venom that Thor dies too... Odin leads the immortals of Asgard and the 43,200 noble warriors reincarnated from Valhalla's 540 doors onto the final battlefield. (These are some very specific numbers that show up all over the world to represent the end of half a precession cycle. There are 43,200 seconds in half a day, the northern half of the Earth is 43,200 times larger than the Great Pyramid, there are 43,200 syllables in the Rig Veda...) Odin is killed by the wolf Fenrir, but his son Vidar kills the wolf. Stars fall, and darkness covers the world. Surt flings fire over the world, and the great tree Yggdrasil burns and falls. The seas flow over the land, and the sun and moon are dead. Yet after all this, "The earth rises up from the sea again... children of men will be found safe... and strange to say, the sun, before being devoured by Fenrir, will have borne a daughter."[279]

American Myth

With the idea in mind that mythmakers who know the facts historically create simple stories to describe ancient events of world destruction and re-creation – and that I see some evidence pointing to the winter solstice in December for what may be the time of the next pole shift – let's not overlook the most modern myth that contains many of the signs we should recognize. Who is associated with judgment over all mankind, with a tree that represents the world axis with a pole star at the top, who is stationed at the North Pole but leaves his position there in late December, and who rides a sky-chariot with eight helpers like a sun god with eight planets? Oh wait, in 1930 a ninth planet (Pluto) was discovered – and by the end of the 1930s, Rudolph was created as an additional 9th reindeer. You can practically rearrange the sounds in the two words as if to hint that eight planets plus Pluto R phludo plus eight reindeer. (Odin also rode a sky chariot at the winter solstice, and in Sweden, the iconic Christmas icon is still depicted as the thunder-god Thor and his elves...) Just meaningless coincidences, I'm sure.

Whose name is so similar to the arch-rival of God who also has little pointy-eared helpers? A name similar to someone who prophecy tells us will lead a rebellion that causes a war in heaven and will be defeated with his angels that will be imprisoned in an abyss at when the present world as we know it comes to an end? In Isaiah 14:13 Satan says: "I will ascend to heaven; I will raise my throne above the stars of God, and I will sit on the mount of assembly in the recesses of the north." Santa descends from the North Pole into our homes like a thief in the night at midnight, through stealth, through the vertical axis of the chimney, never normally through the door. Much like Jesus is supposed to return like a thief in the night – but only to those who are not awake – (otherwise, just like the rounds of the Temple's high priest, he comes on schedule as expected by those who are awake to his schedule) Santa's arrival, though much like a thief sneaking in, is expected at a scheduled time after the usual interval in the astronomical cycle, but the children who expect him are asleep and miss his coming. Seen from this perspective, maybe even the story of Santa should be considered as one linked to a pole shift.

Santa is also linked to the Sumerian and Babylonian shepherd-god of agriculture and fertility, Tammuz. His role was like the Egyptian Osiris and the Greek Adonis, including an unfortunate death and attempts at resurrection. Rituals about Tammuz could be linked to memories of a pole shift, and a dead and reborn sun, possibly at the time of the winter solstice in late December. The Arabic writer Al-Nadim, in his 10th century book: <u>Kitab al-Fehrest</u> said that in tradition, Tammuz was killed prematurely and his murderer "ground his bones in a mill and scattered them." These references to a mill and scattering are appropriate for a "god" associated with a pole shift. To symbolize the untimely death of a god (the sun?) that should have been immortal, an evergreen tree would be cut and used much like a modern Christmas tree. Jeremiah 10:1-4 warns us about living in fear of heavenly calamities and worshiping Tammuz or lamenting his death this way:

"Do not learn the ways of the nations, and do not be terrified by the signs of the heavens. Although the nations are terrified by them; for the customs of the peoples are delusion; because it is wood cut from the forest, the work of the hands of a craftsman with a cutting tool. They decorate it with silver and gold, they fasten it with nails and hammers so it will not totter." The Bible warns us not to do have such a "Christmas tree," to not mourn or celebrate Tammuz or his wife Ishtar, the Babylonian "queen of heaven" - their goddess of prostitution, and mother of harlots. "But the kids love Christmas! What's the harm in having a tree to decorate or leaving milk and cookies out for Santa?" Jeremiah 7:18 "The women knead dough to make cakes for the queen of heaven; and they pour out drink offerings to other gods in order to spite Me." At a holiday allegedly celebrating the birth of Jesus, (which clearly occurred in late September) modern Christians spend way more time adoring a substitute deity of very questionable origins and focus on materialism…

Santa is not the only "modern" story with potential pole shift details we so often overlook…

The first known references to a story similar to "Chicken Little" come to us from approximately Plato's era, documented in Buddhism's <u>Daddabha Jakata</u>, in which a rabbit disturbed by a falling fruit thinks the world is coming to an end. In our modern English version, Chicken Little has an acorn fall on her head, jumps to the hysterical conclusion that "the sky is falling," gathers up other animals encountered on her journey of warning everyone, and they are all eaten by the fox in the end. The lessons we are expected to learn from this story are to not overreact to minor things (like an acorn) and also to investigate before believing everything we are told.

But should we ignore this well established ancient story in which characters believe the sky is falling and the world is ending – when there is evidence that pole shifts have occurred in the distant past, and that one very good way to describe the sudden new movement of the Earth's surface and the strange disappearance of stars below the horizon – would be to say that the sky fell? The Greeks, the Chinese, the Aztecs, the Easter Islanders, the Celts, the Lapps, the Maya, Tibetans, and many tribes throughout Africa and North America all described the sky falling at the end of a previous world. When the Bible describes the next end of the (present) world, Revelation 6:13 says: "and the stars of the sky fell to the earth." Mark 13:25 warns: "and the stars will be falling from heaven." Perhaps our ancient ancestors knew that the sky did fall, and we should be aware that someday it will fall again during a world (as we know it) ending event.

"A cosmic event of the first order can easily be overlooked when it hides modestly in a fairy tale."[280]

Hey Diddle Diddle,
The cat and the fiddle,
The cow jumped over the moon;
The little dog laughed
To see such a start,
And the dish ran away with the spoon.

This innocent sounding nursery rhyme can be traced back through 16th century England all the way to ancient Persia. And it might not be a coincidence that every reference in it has an ASTRONOMICAL counterpart. I think it could be an ancient description of the last pole shift. What did our ancient ancestors view in the sky that made them record this event in a way it would be re-told and passed on for thousands of years?

In the 1569 writing "The Life of Cambeses King of Persia" there is mention of "dance called hey-diddle-diddle."

The diddle is the dance in this case, of the astronomical bodies moving across the night sky.

The constellation Leo is the cat. No surprise this would be mentioned, for it was approximately 12,900 years ago in the zodiacal "Age of Leo" when the last catastrophic pole shift occurred and the "Ice Age" in North America ended with the Pleistocene Extinction. The "Ice Age" in America ended because the North Pole, which had been centered near Hudson Bay, moved suddenly to its present location.

The constellation Lyra – the lyre – the fiddle – is the fiddle of the diddle. It contains the star Vega, the 5th brightest star in the sky. Vega was our pole star 12,000 years ago, just after the pole shift, but long before the precession of the Earth's axis, (which moves across the sky in a circle every 25,900 years or so) had moved enough to point our axis at Polaris, our current pole star.

The constellation Taurus, the bull (the cow) appeared to rotate around (jump over) the Moon, which would only happen if the surface of the Earth itself was rotating quickly around a new and unusual pivot point during a pole shift.

Astronomically, the little dog is the constellation Canis Minor, which includes the star Procyon, our 7th brightest star. This constellation was looking right at the new and strange pivot point in the sky, near the Moon and Taurus (and Gemini.) If the direction looked at by the little dog (Canis Minor) was not crucial to the story, we surely would be told about the big dog next to it – Canis Major – with the "Dog Star" Sirius, the BRIGHTEST star in the sky, never mentioned in this rhyme. Because only the little dog sees/laughs/jumps when he looks right at the crucial part of the sky, we could assume this clue is important astronomically.

But the phrase "little dog" has another meaning, lost in time and translation of language... Jesus used this phrase with meaning in Matthew 15:26. A bit earlier in Matthew 15:16, Jesus chastises his followers for being "lacking in understanding" because learned Jews should have known their scriptures well enough to understand Him. The Greek phrase "kunarion" means "little dogs," and in 15:26 Jesus is outside Israel, near Tyre and Sidon, when a Canaanite woman asks Him for help, and His first response to her is to say "It is not good to take the children's bread and throw it to the little dogs." He meant that His teachings were for the Jews first, and not yet for other people not "chosen" or trained by their scriptures to have a chance of understanding the message. In our nursery rhyme, saying that the little dog laughed to see such a start might mean that most of the world's people, untrained in science and astronomy, clueless about the cycles of destruction, didn't understand what was happening and didn't know what to make of the changes they witnessed during the pole shift.

The laughter may also be important to the story. In many myths of the transition between world-destruction and re-creation, the sun-god hides in a cave and refuses to come out until another god make them laugh. The Egyptian sun-god Ra had retired and refused to carry on his solar duties until Isis made him laugh. The Japanese sun-goddess Ameratsu is similarly enticed to resume her solar journey. The Cherokee's Mother Sun is laughterless over the death of her daughter, much as the Greek goddess Demeter was laughterless over the death of her daughter Persephone. Laughter may be symbolic of the joy of the sun's return to life and the start of a new world. Laughter may also be similar to the vibrating undulations of a shaking world, as layers of crust grind and vibrate beneath us.

The constellation Crater (the dish), known to Babylonians as a sign of death and as a gate to the underworld – is a dish.

The constellation Ursa Major, the Big Dipper, is the spoon. It contains Polaris, our current pole star. Both the dish and the spoon "ran away" when the pole shift made them appear to suddenly move about 30 degrees from their normal course in the sky. Of course this is just a theory. Despite all the fairly clear astronomical references in the nursery rhyme, there is no proof it refers to an astronomical event.

Myths and legends from around the world yield tantalizing clues of a former pole shift. We are told there was a cataclysm in which humanity was judged by the gods and most people were killed. Lands sank, seas rose, and stars fell in conjunction with a great flood. An ancient paradise, usually an original island homeland, had to be abandoned. Mythical beings act out what are clearly astronomical displays in the heavens. The great world mill, with its seemingly endless rotation, is broken and destroyed, and a new mill with a new rotation has to be created. In the end, a new order for the world is reestablished.

The Ancient Island Homeland Was Destroyed

Egyptian records in the Edfu Texts claim their "gods" were from the Ta-Neteru – the land of the gods – a large island south of Egypt known as "the Homeland of the Primeval Ones... the texts are adamant that the agency that destroyed this island was a flood. They also tell us that it came to its end suddenly and that the majority of its 'divine inhabitants' were drowned."[281] Sumerian tablets say their gods fled from a mountainous island in the south, across the Indian Ocean, called Dilmun.[282]

In the ancient Persian Zend-Avesta, Ahura Mazda warns Yima, the first king of men, of an upcoming "dire winter, which is to destroy every living creature by covering the land with a thick sheet of ice" over the Airyana Vaejo, the original Paradise of the Iranians.[283] An excerpt from the Airyana Vaejo speaking from a later point in time states: "...At that time Airyana Vaejo had a pleasant climate, with seven months of summer and only five months of winter. The forests were rich with game and the fields with grains. In the valleys many brooks flowed. This land however turned into a cursed land, where for ten months it was winter and only two months was it summer, following the attacks of Ahra-Majnyu." This was an evil power they blamed for the climactic changes that made them flee their original homeland after it shifted to Arctic or Antarctic conditions.

In this extract from a similar Persian story in the Zoroastrian Vendidad, Ahura Mazda commands Yima to build a shelter:

"And Ahura Mazda spake unto Yima, saying: 'O fair Yima, son of Vîvanghat! Upon the material world the fatal winters are going to fall, that shall bring the fierce, foul frost; upon the material world the fatal winters are going to fall, that shall make snow-flakes fall thick, even an aredvî (large unit of measure) deep on the highest tops of mountains. And all the three sorts of beasts shall perish, those that live in the wilderness, and those that live on the tops of the mountains, and those that live in the bosom of the dale, under the shelter of stables.... Therefore make thee a Vara (enclosure), long as a riding-ground on every side of the square, and thither bring the seeds of sheep and oxen, of men, of dogs, of birds… bring the seeds of men and women, of the greatest, best, and finest kinds on this earth; thither thou shalt bring the seeds of every kind of cattle, of the greatest, best, and finest kinds on this earth.... All those seeds shalt thou bring, two of every kind, to be kept inexhaustible there, so long as those men shall stay in the Vara."[284]

This sounds a lot like the story of Noah's Ark, except that the problem is not from a flood but from many "fatal winters" – a multi-year (or permanent) climate change. It also offers the tantalizing hint of high technology capable of storing the genetic "seeds" of men, women, and animals. With 21st century technology, if we were faced with maintaining genetic diversity through a catastrophe, we would undoubtedly store DNA samples and frozen embryos – and not necessarily attempt to maintain live herds of huge animals. Perhaps the ancient homeland of Airyana Vaejo was Atlantis. Vimana "arks" built by them may have had more advanced technology than we imagine.

In ancient Mexico, the only survivors of the great flood were Coxcox and his wife Xochiquetzal. Safe in the hollow of a floating cypress tree trunk, they are said to have floated to the top of a mountain near the first Toltec city, founded by King Mixcoatl. In pre-Columbian Mexico, the syllable/sound "atl" meant water, and the suffix "-atlan" additionally meant "near the place of abundance." Many towns near the coast like Mazatlan still have names including this syllable.

The ancient Mexican sea gods were named Atlahua and Atlanteotl. The Aztecs of central Mexico told the Spanish conquistador Hernan Cortes that their ancestors came from a white island across the sea called Atzlan, a bright land of shining light and whiteness that used to contain seven cities near a sacred mountain.[285] The Aztec capital of Tenochtitlan (now Mexico City) was "built on a lake-island surrounded by concentric canals, specifically because this arrangement reflected the original topography of Aztlan, the homeland of their ancient traditions."[286] Plato described Atlantis the same way, with rings of land and water surrounding the capital city. Cortes' secretary and biographer, Francisco Lopez de Gomara, wrote in 1552 that the Aztec homeland was apparently the same land as Plato's Atlantis.

It is also interesting when we consider the linguistic similarity: "The consonantal combination tl is exceedingly rare in most European languages, including Greek... but among the rare exceptions to this are the words Atlas, Atlantic, and Atlantis. In Nahua and related Central American languages, however, the tl combination is exceedingly frequent – perplexingly so for European tongues. This may suggest that the source of the word Atlantis may have been in close linguistic contact with Central and South America."[287] Both the Aztec stories, and the Egyptian stories told to Plato, may have had shared roots in Atlantis. Even in the Arctic, the Inuit of Baffin Island remind me of Atlantis when they tell of the similarly named Adlivun: "The Earth opened beneath them and all of them fell deep down to the land of Adlivun, which is the land beneath the land and the sea."[288]

The Okanagan Indians of Western Canada, despite living in the Rocky Mountains, associated a great flood with an island "far off in the middle of the ocean." The Cherokee, in the Appalachian Mountains, also suggest the great flood had something to do with the "great island floating in a sea" and that "There is another world under this one, and it is like ours in everything – animals, plants, and people – save that the seasons are different."[289] The Inca of Peru and Bolivia said the gods had brought agriculture to the lands near Lake Titicaca when they came "out of the regions of the south immediately after the deluge." The Inca also had a ritual every winter solstice at Macchu Picchu in which a magic cord was tied to a giant pillar to guide the sun and prevent it from losing its way across the sky.[290] Were they worried because, on some previous solstice, the sun had gotten lost and wandered through the sky?

India's first leader using passive resistance against the British, the man Mahatma Gandhi called Lokamaya or "Beloved Leader of the People" was named Bal Gangadhar Tilak. Tilak's books on India's ancient Sanskrit Vedas earned him great respect from British scholars at Oxford and Cambridge, who asked that he be released from prison. In 1903 he wrote a book about the original homeland of the people and said: "It was the advent of the Ice Age that destroyed the mild climate of the original home and covered it into an ice-bound land unfit for the habitation of man."[291] And "Ancient India's Mahabharata describes an ancient homeland of the gods where 'The day and night together are equal to a year.' In the Surya Siddhantha they clarify that many days elapse between the first light of dawn and the eventual appearance of the sunrise. Antarctica fits these descriptions well."[292]

Deity Sacrifice

Aztecs and many other people mistakenly believed they needed to prevent another catastrophe with a steady stream of human sacrifices. These bloody rituals may have originated in much more ancient ideas of personal purification involving the sacrifice of the will – one's own egotistical desires. But they degenerated into attempts to placate a seemingly bloodthirsty sun-god who had briefly stopped functioning normally and had only returned to his duties after billions of people were killed. Many cultures in the Americas believed that regular human sacrifices were necessary at certain astronomically important times to "keep the sun moving."[293] In the Aztec legend, the sun went dark at the end of the last age, and two gods sacrificed themselves to be reborn as the new sun (and moon.) But "Tonatiuh demanded the fealty and blood of the other gods before he would get back to work travelling along the ecliptic... one by one [other gods] gave themselves up to have their hearts removed by Quetzalcoatl. Fortified by this blood offering Tonitiuh became Nahui Ollin, the 'Sun of Motion.'"[294] Humans were expected to continue the tradition and satisfy the sun's appetite for blood lest it fail to light the sky or move through it anymore.

This seems a bit strange to modern Westerners, who have generally been taught that Christ sacrificed Himself for our salvation, but "this is in keeping with the Mesoamerican idea that deity sacrifice must take place before world renewal."[295] In their cyclic worldview there is always the threat of future catastrophes and future rebirthing of world ages requiring additional sacrifices. The Mayan deity Nanahuatzin "hurled himself into a huge fire and became by that act the sun-bearer of this last and present age. But that single sacrifice – even though of a god – was not enough. To keep the sun moving, continuing sacrifices are required... [including] the gathering of victims..."[296]

In the Mayan epic the Popul Vuh, a pair of characters known as "the Hero Twins facilitate the resurrection of their father" One Hunahpu, who is the winter solstice sun. He dies at the crossing point of the Milky Way and the ecliptic – the galactic center throne. The Hero Twins could represent the North and South Poles or Venus and Jupiter. "One Hunahpu's resurrection signals the beginning of a new World Age. Since the December solstice sun is what gets reborn, we are drawn to recognize the end-date alignment" which along with the Pyramid of Kukulkan at Chichen Itza, "is a precessional alarm clock with its alarm set for the twenty-first century."[297] The Maya sun god of the winter solstice is the sacrificed deity who helps usher in the new world age.

In Aztec mythology, before the first sun had risen at the creation of the fifth age, "the gods assembled themselves at the ceremonial city... they performed penances around a sacred fire, symbol of the divine center.... It soon became clear that a very great sacrifice would be necessary.... In a blissful moment of self-sacrifice, two of the deities hurled themselves into the fire.... Thus the world was made through the self-sacrifice of two deities, through offerings to the central axis mundi."[298]

"In the ancient world there was a very widespread belief in the sufferings and deaths of gods as being beneficial to man. Adonis, Attis, Dionysos, Herakles, Mithra, Osiris, and other deities, were all savior-gods whose deaths were regarded as sacrifices made on behalf of mankind; and it is to be noted that in almost every case there is clear evidence that the god sacrificed himself to himself."[299]

Jesus Christ sacrifices Himself on a cross reminiscent of the sun on the crossing point of the ecliptic and the Milky Way. The Norse god Odin hung himself on a tree to sacrifice himself at the end of a world for the beginning of a new one, and we are directed to consider Odin's Eye at Mimir's Well of Wisdom, presumably the sun at the galactic center. In Egyptian tales the son of the highest god, Horus sacrifices his eye to help resurrect his father, the great god Osiris – who in some texts, is dead for three days.[300] In ancient Persia, Mithra sacrificed himself for the benefit of the universe, as Vishnu repeatedly does in Indian literature.

At one central location, perhaps symbolic of the galactic center, Adam made an altar for sacrifices, where Cain and Abel made theirs, which was eventually used by Noah, and Abraham, and the Temple of Solomon... But aside from verifying a trend involving a sacrificed sun-god to restart the world, and the idea that continuing sacrifice is required for our sins... these stories fail to tell us much about timing – except that the solstice sun god deity, often the son of the galactic center supreme deity – usually sacrifices himself on the cross, or the crossing point of the ecliptic and the Milky Way.

Manly Hall wrote over a century ago in The Secret Teachings of All Ages that as part of being initiated into the mysteries, "the candidate, after being crucified upon the cross of the solstices and the equinoxes, was buried..."[301] One French author a century before that said "Jesus was only one of the solar saviors 'crucified' on the cross of the solstices and equinoxes."[302] The divine solar solstice sacrifice is always made at this crossing point. Because of Earth's cycle of precession, this does clue us in to expect such events during small windows in time – like the early 21st century.

The Unhinging of the Mill

One of the best and most comprehensive books ever written on the analysis of world mythology from an astronomical perspective is Hamlet's Mill: An Essay Investigating the Origins of Human Knowledge and its Transmission Through Myth. Shakespeare's story of Hamlet may be the most well-known version of a certain theme of myth in which a golden age ruled over by a good king ends, followed by bad times and a quest of the hero who is the son of the good king to reclaim the throne from his bad uncle and restore a golden age. Every culture has such a story, at least as far back as Egypt's god-king Osiris, his evil brother Set, and his son Horus. Hamlet is actually one of the newest versions, and therefore has gone through many millennia of translations and has been watered down in meaning.

Many key references, often no longer understood by the storytellers or translators, have been edited out of such stories. But some ancient clues, fortunately, have "tradition preserved intact even if half-understood."[303] Enough remains among many similar stories from around the world for the authors of <u>Hamlet's Mill</u> to make some amazing conclusions, including: "The Mill is thus not only very great and ancient, but it must also be central to the original Hamlet story." Also: "the unhinging of the Mill is caused by the shifting of the world axis."[304]

Iceland's most ancient skaldic tale involves a mythical king named Frodhi. In more recent versions, influenced by Christianity, Frodhi is a good human king ruling during a golden age at the time of Christ and the Roman Emperor Augustus. "But his name is really an alias of Freyr, one of the great Vanir or Titans of Norse myth."[305] The end of his rule is the end of a world age linked to a cosmic war between pantheons of gods as is described by many cultures. A collection of ancient Icelandic kenningar – oral stories of many Viking bards – was eventually written down by Snorri in the Skaldskaparmal.

In this oldest Icelandic book, the legends are already ancient. King Frodhi owns a magic mill named Grotte, the crusher. The poles are represented by two giant maidens named Fenja and Menja who are strong enough to turn the king's mill, and they work hard to churn out a golden age for countless years until calamity approaches: "Wake now, Frodhi!... The war news are awake. That is called warning. A host hither hastily approaches to burn the king's lofty dwelling... Now we must grind with all our might. No warmth will we get from the blood of the slain... Now broke the large braces beneath the mill – the iron-bound braces. Let us yet grind..." King Frodhi is killed and a new king from across the sea takes over. King Mysingr loads the mill onto his ship and forces the two giant maidens to keep grinding until the spinning mill sank the ship: "the huge props flew off the bin, the iron rivets burst, the shaft tree shivered, the bin shot down, the massy mill-stone rent in twain."

I am reminded of Matthew chapter 24. Here Jesus describes wars, great tribulation, stars falling from heaven, Noah's flood, and then says in verses 41-42: "Two women will be grinding at the mill, one will be taken, and the other will be left. Therefore be on the alert, for you do not know which day your Lord is coming." Jesus' final warning about His coming includes a sign about two women grinding and an end to their turning the mill. This could definitely refer to a change in the earth's rotation at the time of the great tribulation. Jean Phaure wrote similar comments in Le Cycle de L'Humanitie Adamique, which could be paraphrased as the Aquarian Age seeing "the Second Coming of Christ, the Last Judgment. Then a new cycle of humanity begins, probably with a reversal of the poles."[306]

Centuries before Christ, the ancient Greek prophetess, Sibyll, was known to go into a trance and make the frenzied commentaries we know as the Sibylline Oracles. For our purposes, her most relevant comments are found in Book V, which in the 1918 translation from H.N. Bate says: "And in his anger the immortal God who dwells on high shall hurl from the sky a fiery bolt on the head of the unholy: and SUMMER SHALL CHANGE TO WINTER IN THAT DAY." Bate compared this to a line in Book VIII: "And then the imperishable God who dwells in the sky in anger will cast a lightning bolt from heaven against the power of the impious. INSTEAD OF WINTER THERE WILL BE SUMMER ON THAT DAY." It sounds like our ancestors may have witnessed a cosmic transference of electrical energy to the Earth, an event which would be much more likely if the inner solar system were briefly filled with interstellar dust to aid the flow of electrical currents. This could conceivably recharge the Earth's weakened magnetic field back to levels recorded in the past, and potentially affect the polarity of the Earth's geodynamo.

In Finland, the Kalevala is the unofficial national epic poem, with tens of thousands of verses describing everything from the creation of the world onward. The hero Vainamoinen is born to a virgin goddess fathered by the highest god. In Finland's version the first rune mentions a helper of Vainamoinen named Sampsa who helps Vainamoinen cut down trees. This is noteworthy both because the world axis is represented by a world tree, and because Sampsa is effectively the biblical Samson, whose outlandish exploits end with being forced to grind at a mill and eventually toppling the pillars and killing everyone under the pillars. In the Russian version of these stories, the epic's namesake is Samson Kolyanovic.

But focusing on the Finnish version – the hero-god Vainamoinen tricks "his brother" Ilmarinen, "the great primeval craftsman" who is a smith-god like the Greek Prometheus, to build a powerful magic mill called the Sampo. They first climb a giant pine tree (the world axis?) with the Great Bear (Ursa Major) in its upper branches and they grab the stars... The Sampo mill is eventually forged, other characters cause problems, and the two brothers decide to take the Sampo away by boat: "Come thou here to take the Sampo... turned himself, and pushed against it, on the ground his knees down-pressing, but he could not move the Sampo..." Are we to understand that even "gods" and "titans" have a hard time moving the magic mill that represents the polar axis of rotation?

With an additional helper, and newly acquired magic, and the help of an enormous bull (Taurus?) they finally "carried forth the mighty Sampo... to the boat away they bore it." There is a magic battle with other gods led by Louhi, in which the boat is sunk "and the Sampo broke to pieces." The world axis is ruined. There is a pole shift. Vainamoinen wants to rebuild but how can he: "Whence shall now the screws be fashioned? Whence shall come the pegs to suit me?"

Vainamoinen's mill peg is almost definitely the pole star. Since he climbed the tree axis and found Ursa Major in its branches, we should note that this constellation is near the celestial North Pole. We should note here the Arabic name for the star beta Ursae Minoris is Kochab, which means "mill peg." Ludwig Ideler commented on the Arabic name in 1809 and wrote: "The sphere of heaven was imagined as a turning millstone, and the North Pole as the axle bearing in which the mill-iron turns." Vainamoinen has lost control of the axis of rotation as his golden age is ending.

Finland's Kalevala continues to explain that when the Sampo mill broke apart: "the moon came from his dwelling, standing on a crooked birch tree, and the sun came from his castle, sitting on a fir tree's summit." The sun and moon move from their normal paths during the pole shift. There are references to the world tree axis and to a crooked or bent tree axis. Could Vainamoinen's original story be 13,000 years old, when the winter solstice sun was at the horns of Taurus near the galactic anti-center? That time period would have been considered the Age of Leo, defined by the position of the spring equinox sun in the lion constellation of Leo. We have already noted the link to the biblical Samson, and we know Samson tearing apart the lion refers to the much earlier Age of Leo...

In the Finnish epic, Vainamoinen sings as he tries to fix everything, but the golden world age is ending, he is dying, and the uninitiated don't even understand his description of a pole shift as he: "Sang the songs of by-gone ages, hidden words of ancient wisdom, songs which all the children sang not, all beyond men's comprehension, in these ages of misfortune, when the race is near its ending.... Men will look for me and miss me, to construct another Sampo." The authors of Hamlet's Mill comment: "As for the name Sampo, it resisted the efforts of linguists, until it was found that the word derived from the Sanskrit skambha, pillar, pole. Because it grinds, Sampo is obviously a mill. But the mill tree is also the world axis..."[307]

The authors of Hamlet's Mill ask: "Why does it always happen that this Mill, the peg of which is Polaris, had to be wrecked or unhinged? Once the archaic mind grasped the forever-enduring rotation, what caused it to think that the axle jumps out of the hole? What memory of catastrophic events has created this story of destruction? Why should Vainamoinen (and he is not the only one) state explicitly that another Mill has to be constructed? Why had Dhruva to be appointed to play Pole Star – and for a given cycle? ...There is quite a collection of myths to show that once upon a time it was realized that the sphere of fixed stars is not meant to circle around the same peg forever and ever. Several myths tell how Polaris is shot down."[308]

Despite mentioning catastrophes and floods and the end of a world, they conclude merely that our ancient ancestors knew of the precession cycle of the equinoxes. I must disagree; there is too much evidence for geological catastrophe and mass extinction, and too many mythical details of heavenly wars between pantheons of gods fighting for control during chaos with earthquakes and a misbehaving sun. These details would not be necessary to describe a slow, gradual, uneventful process like mere precession, barely observable over a 26,000 year cycle without catastrophes...

Avatars of Destruction

In ancient India, Shiva is the destroyer. The high trinity or trimurti of the Hindu religion has Brahma as the top god, with Shiva repeatedly coming to destroy the world, and Vishnu repeatedly coming to re-create it and restore cosmic order. The roots of their names, "SH-IV" and "VI-SH" are mirror images, indicative of their opposing but related roles. Shiva is not viewed as an evil destroyer but as the destroyer of worlds gone evil themselves. Shiva creates, transforms, and protects by destroying evil and purifying the world. Vishnu, the preserver, is viewed as a protecting preserver of good. Shiva wears a serpent around his neck. Vishnu reclines while resting on the coils of the serpent. Both Shiva and Vishnu are often depicted with blue skin, perhaps chosen because it is the color of light from the galactic center?

The Bhagavad Gita (4.7-8) describes the typical role of Vishnu's avatars:

Whenever righteousness wanes and unrighteousness increases I send myself forth.
For the protection of the good and the destruction of evil,
And for the establishment of righteousness,
I come into being age after age.

One of Vishnu's early avatars is sometimes described as a fish-man, but more often just as a fish (specifically a carp) named Matsya. In this form Vishnu saved the world from a cosmic flood and protected the ancient knowledge in the Vedas by warning the human king, Manu, of an impending global flood and advising him to build an ark. As the tortoise Kurma, Vishnu supported the world during a cosmic war where the Milky Way was involved with churning the axis and producing poison. As the boar Varaha, Vishnu took the earth goddess as his bride and saved her/Earth from the depths of the ocean. There are many other ancient avatars of Vishnu, and one in future prophecy, in which Vishnu is expected to return as Kalki.

Kalki is the final incarnation of Vishnu, and is expected to come "like a comet" carrying a fiery sword to annihilate the wicked. Because "it is Vishnu's function to return as avenger at fixed intervals of time..."[309] some claim to know when Vishnu will come back as Kalki. Kalki arrives in the form of a man on a winged white horse at the end of the current Kali Yuga Age, to renew and restart the world in the next age. This future avatar of Vishnu sounds a lot like the Return of Christ on a heavenly white horse in Revelation 19:11.

Shiva also returns under many avatars. Legends say he was born as Piplaad to the sage Dhatachi, who abandoned him before he was born, due to the unfortunate astronomical position of "Shani" – Greeks associated Shani with Saturday, then Saturn, but Shani may have originally represented another astronomical body. I doubt it was originally Saturn because Shiva cursed "Shani" and made it "fall down from its celestial abode." It would make more sense to me that Shani is a former pole star.

In another incarnation Shiva was involved in a cosmic war with Brahma and took on the form of Bhairava, most often associated with annihilation. As Bhairava, he severed one of Brahma's five heads. Ashwatthama is an avatar associated with anger and death, and with this form of Shiva swallowing a cosmic poison even he could not handle. This is what turned him blue. (Other sources say almost every Hindu god is depicted as blue because they have a blue aura. I point out again that the galactic center lights up blue when it is in its active phase.) This avatar of Shiva was also involved in the (nuclear) war described in the Mahabharata, during which Vishnu prevented Shiva from killing all the people and limited destruction to the oppressors.

In Mayan mythology, Huracan is the god most associated with the creation and destruction of worlds. He is also known as "Heart of the Sky" – potentially linking him to the galactic center – and also "One-Leg," which is a name or description used around the world for a god of the polar axis. Sometimes Huracan's missing leg has been replaced with a serpent. It probably represents the cosmic serpent that was taken away at the creation of the present world, when "the sky fell and the world was flooded."

Many versions of the global flood myth exist in different Mayan regions; in some versions the gods are angry that someone is burning something and "smoking the sky" and mankind is punished – in other versions the cataclysm is not punishment but natural, the gods take pity on humans, and people are warned by a god of the upcoming flood and they prepare a boat. Some local versions say there are a few survivors that use a canoe to reach a mountaintop and repopulate the world, while in other versions humanity is completely destroyed and must be created again. Huracan is credited with participating in humanity's creation – repeatedly having helped to create people in at least three distinct world ages. He repeatedly invoked the earth to rise out of the depths of the seas. But he also sent a great flood or a fire to destroy his human creations several times; at least twice because they were inferior, and once because they had been made too intelligent.

Like the Greek Prometheus, Huracan gave fire/knowledge to mankind. When the gods decided men had gained too much wisdom, he clouded mankind's eyes and minds with mist (cosmic dust?) because their intelligence was becoming a threat to the gods. Huracan plays a large role in the creation and destruction of world ages, and his only other mention is a small role in the Popul Vuh in which he assists the Hero Twins in their efforts to dethrone the false god Seven Macaw from his polar perch before they resurrect their father, the winter solstice sun god One Hunapu. Huracan is also the god of great storms, wind, and lightning, from which Europeans coined the word hurricane.

In ancient Egypt, many different gods are associated with the sun in different ages, including Ra, Osiris, and Horus. The sun often experiences death and resurrection under a new identity when near the stinger of Scorpio by the galactic center. The succession of several sun-gods has made many Egyptologists wonder why the Egyptians felt that the sun would die and need to be reborn with a new identity, and why this always happened with the sun at the galactic center.

As Graham Hancock explains: "we can deduce from the texts that Ra, i.e. the sun's disc, was seen somehow to merge or unite – or 'coalesce' – with Horakti." (Horus of the Horizon) With the astronomical alignment of the galactic center and the solstice sun "the composite deity Ra-Horakti" is created. Only when the two bodies are united together at the throne (pretty clearly indicating an astronomical alignment) are new powers manifested in the double god Ra-Horakti (or perhaps in a previous event, Atum-Ra, and in a later event, Horus-Osiris) We are told that "Osiris remained 'inert, asleep or listless, and completely passive' until the Horus King [solstice sun] was able to undertake a journey to the Duat [a region of the underworld/sky] and 'visit his father' and 'open his mouth' i.e. bring him back to life."[310]

When this happens, the goddess Hathor – both the mother and the daughter of Ra – indicating repeating cycles – is sent by the sun-god Ra, or specifically by his weaponized aspect, the Eye of Ra, to punish and destroy mankind, though she is always tricked into sparing a small portion of humanity. As I quoted many pages back from Egyptian mythology on this subject: "Of itself the [Ra's] Eye is not strong enough to destroy them. Let it descend upon them as Hathor. So that goddess [Hathor, always drawn with the sun disk on her head] came and slew mankind."

Ancient Egypt's Coffin Texts offer another angle on this astronomical alignment of the solstice sun and the galactic center eye causing great destruction. Spell 3:16 says: "I am the all-seeing Eye of Horus, whose appearance strikes terror, [along with Hathor] Lady of Slaughter, Mighty One of Frightfulness, who takes the form of blazing light, whose appearance Ra ordained, whose birth Atum [the sun-disk] established... all mankind will cringe beneath your might, they will respect you when they behold you in that vigorous form... Behold it (the Eye) will be stronger than all the gods, It has mastered the dwellers at the ends of the Earth, (the poles?) it is sovereign over every god... No one will come who can withstand me, except Atum, for it was he who originally moved me and put me before him so that I could wield power." The galactic center eye appears in alignment with the sun and nearly destroys mankind. Though associated with the color red for the blood of most of mankind being spilled during her wrath, Hathor's sacred color is turquoise blue.

There is a well-known Egyptian zodiac, considered so important by Europeans [Masons?] that the original stone zodiac was removed from Hathor's Temple at Dendera and has been at the Louvre for over fifty years. A replica of the Dendera Zodiac was created for the original location. Strangely, the North-South axis through the stars of this zodiac does not represent how the sky looked at the time the temple was built. It represents the sky at the start of the Age of Virgo, when the spring equinox sun rose in Virgo approximately halfway back through the almost 26,000 year cycle of precession. Dr. Paul LaViolette believes it points to a date about 13,000 years ago and says: "Atum sent out Hathor from the center of the universe (the Galactic center) to cause a mass destruction of mankind, a catastrophe that was terminated by a global flood, so it seems quite fitting that the Dendera Zodiac refers to this same prehistoric cosmic event."[311]

John Lash wrote about the Dendera Zodiac in Skies of Memory, and summarized his conclusions for Colin Wilson and Rand Flem-Ath to use in their own book: The Atlantis Blueprint. Nash noted that unlike most Egyptian temples, the Dendera temple had foundations 20 feet below the surface, indicating great age for the original building. More importantly, Nash concluded that the most important axis of the Dendera Zodiac "extended to Virgo's feet marks the tail of Leo at a point that corresponds by precession to 10,500 BC. All in all, axis E marks the moment of precession when one full cycle ends and a new one begins. We are currently living through the last two centuries of the full 26,000 year cycle… Whoever designed Dendera was looking ahead in time to our age, when the spring equinox occurs under the tail of the western fish, because this is the time when the entire cycle culminates."[312] Hathor's next act of destruction may be due very soon, when the spring equinox is in Pisces and the winter solstice sun aligns with the galactic center.

Phaethon

The story of Phaethon may be the best example of a description of a pole shift in ancient mythology. Ovid, Nonnos, and other ancient Greek writers recorded oral traditions and explained that when the sun-god Helios met his half-mortal son Phaethon for the first time, he took an oath by the River Styx that he would grant his son any wish. Phaethon asked to drive his father's sun-chariot for a day. Helios argued against it, and gave many warnings of what dangers and obstacles there are in the sky, but Phaethon insisted and Helios had taken an oath. As expected, Phaethon could not handle the horses' reins, and as Ovid tells us, Phaethon drew back in fear by the stinger of Scorpius, and let go of the reins. The River Styx represents the Milky Way and the stinger of Scorpio points at the galactic center where so many sun gods lose their way and temporarily die.

The sun wanders through the sky and burns the Earth: "Great cities perish, together with their fortifications, and the flames turn whole nations into ashes" [while Atlas] "fails to balance the world's hot axis on his shoulders."

Nonnos wrote that "the very axle bent which runs through the middle of the revolving heavens." (Sounds like a pole shift to me.) The world is almost completely destroyed, but Zeus comes to the rescue, shooting Phaethon with a bolt of lightning to end his crazy ride. Zeus coming to the rescue might mean that Jupiter moved in to join the astronomical conjunction of galactic center and solstice sun (though Dr. LaViolette suggests that "Zeus signifies the galactic core.")[313]

It is possible that we should view Phaethon not as the sun or sun chariot but as a temporary rider, i.e. an astronomical body that briefly tried to ride the sun or briefly experienced a conjunction with the sun at the time of the pole shift. In End Times and 2019 I explained why (as explained to me by astrophysicist Dr. Gary Vezzoli) the Earth can experience a gravitational jerk when a large planet appears to move in or out behind the corona of the sun. The extreme heat of the corona seems to scatter the muon neutrinos (gravitons) that seem to convey the force of gravity. Jupiter will appear to be hovering at the corona of the sun on December 28, 2019, at the end of the wedding in the sky I have described, and if a pole shift ever happens during such an alignment, Jupiter's repeated movement behind the sun's corona could play a role.

During the last pole shift when the winter solstice sun was at the horns of Taurus near the galactic anti-center, ancient India's writings suggest that Saturn played this role. The Surya-Siddhanta (V 8:13) says that "in Taurus, the 17th degree, a planet... will split the wain of Rohini." The Pancatantra (V 238-241) says that "When Saturn splits the wain of Rohini here in the world... the earth, having as it were committed a sin, performs, in a manner, her surface being strewn with ashes and bones... a sea of misfortune, destruction befalls the world... men wander recklessly about, deprived of shelter, eating the cooked flesh of children..." all because of "the bad experiences Saturn had with Auriga's vehicle"[314] near beta zeta Tauri and the Hyades (in Taurus.) This sounds like Saturn may have played the role of the rider on the sun (Phaethon) and Auriga's vehicle may refer to the sun-chariot.

Greek myth has Phaethon's watery grave in the stream of Eridanus, from the Babylonian Eridu, the mouth of the rivers. But the river for Phaethon is not in Mesopotamia's Tigris or Euphrates, but in the Milky Way, where he is often drawn on ancient sky maps near Orion's foot. Some scholars, like Richard Allen and W.H.D. Rouse, flat out say that Eridanus and the Milky Way are one and the same. I have no doubt that the sun's wild ride ended when it was in the Milky Way, probably recurring every 12,900 years or so when the solstice sun (winter or summer) is at the Milky Way near the galactic center or anti-center.

A similar myth exists in the legends of British Columbia's Bella Coola Indians. A human woman married the sun-god, and they had a baby boy who eventually asked his father to let him carry the sun in its path across the sky. His father told him he lacked the experience to do it, but after much insistence, the sun-father relented, warning him to carefully follow instructions on how to burn a few small torches early and late in the day, and only burn lots of big bright torches midday. Of course the boy was impatient and quickly burned them all at once. The sun flared up and scorched the world. Many other North American tribes describe an event involving a global fire and flood, with the destruction of almost all people except for a few survivors who often hid underground in caves or went to live with ant-people underground.

Another interesting myth from western Canada comes from the Dogrib and Slavey tribes from Alberta. No one wrecks their father's sun chariot in this one, but it is still worth mentioning here… It begins: "One night the darkness was very thick and snow began to fall… The night continued for so long that it seemed never to have an end. The snow became deeper and deeper… and a decision was made to send messengers to the Sky People to find out what was causing this long night and the deep snow. [Animals were sent to] the Sky World and passed through the trapdoor… [where] there stood a great lodge… the home of Black Bear, [which had] …five curious bags hanging from the overhead rafters… [the animals eventually discovered they] contained the sun, moon, and stars. These they threw down through the trapdoor.

While they watched, the snow began to melt from the heat of the sun. But the snow melted so quickly that the earth was covered with water…. When the flood waters had gone, the peaceful and friendly life on Great Slave Lake was no more. The birds and fish and beasts chose different places to live and soon they forgot the language they once shared."[315] This myth could mean that a dust cloud bagged the lights of the heavens in darkness, leading to long darkness with cold and snow, followed by the heat of the returning sun and a flood of the world, after which people spread out to find new homelands and lost their original shared language – much like the Tower of Babel story. (I often wonder if the shared language was merely a dominant language of the previous civilization, as English became prevalent around the world by the 20th century – or was that lost common language advanced math and science?)

The Egyptians described Phaethon's fall as a "lost eye" whose falling made the Nile hide its source. The sun itself could easily be the lost eye, especially if there were three days of darkness. Of course the lost eye could also refer to the Eye of Horus/Eye of Ra, the blue oval with a bright galactic center pupil that would be visible for about a thousand years after a galactic superwave event. If my theory is correct, the last such event caused a pole shift and the climate of northeastern Africa changed from swamps and lakes and verdant grassland to the current Sahara desert – a very real obstacle discouraging anyone near the Nile Delta from travelling south to find the river's source.

I would also like to point out that the stinger of Scorpius is near the galactic center and points to it. So when the Sun (Helios) is near the river (Milky Way) a bad chain of events leads to the sun veering off course and crashing after reaching the stinger of Scorpio near the galactic center. This is the same region in the sky where the Egyptian sun god Ra was bitten by the scorpion, and in other Egyptian stories it is where the sun god Horus is stung by the scorpion. The sun boat stands still in the sky and the wisdom god Thoth recites a spell: "Back, O Poison!... The boat of Ra will stand still and the sun remain in its place of yesterday until Horus recovers... Down to earth, O Poison! So that all hearts rejoice and the light of the sun go round once more..."

Another legend says that when Ra had grown old, and his hair was like lapis lazuli (blue) Ra heard rumors of mankind's upcoming rebellion against him, and that he gathered the gods in secret to judge mankind. "Let them all come hither in secret so that men may not behold them, and fearing, take sudden flight. Let all the gods assemble in my great temple at Heliopolis" (the city of the Sun.) "At that moment Ra removed himself from the sight of the gods in his boat, and the Throne in the Boat of Millions of Years had no occupant." Ra took "an oath to part with his two eyes, that is, the Sun and the Moon... [Ra revealed his secret name to Isis...] Ra thus became to all intents and purposes a dead god. Then Isis, strong in the power of her spells, said: 'Flow, poison, come out of Ra. Eye of Horus, come out of Ra, and shine outside his mouth." Is this Egypt's explanation of the sun going dark and the end of the normal rotation of the Earth and the appearance of the bright blue light at the galactic center when the sun was near the stinger of Scorpio at the last pole shift? Was the poison that had to come down the acidic cosmic dust that had entered our solar system?

Another interesting zodiac comes from the floor of the Beth Alpha Synagogue from the Galilee region of Israel. The mosaic tile zodiac shows the spring equinox at the cusp of Virgo and Leo, approximately 12,650 years ago. Joseph Campbell, the most respected analyst of mythology, noted that this alignment was way off from an accurate depiction of the sky at the time the synagogue was built around the 6th century A.D. – and he assumed that the builders were simply too ignorant to get it right. "The problem of the religious significance of Helios, the zodiac, and the seasons to a Jewish community of that time is not easy to resolve."[316] But it seems much more likely they were very aware of the astronomy and were showcasing the significance of that ancient time frame. Campbell was also confused about the Beth Alpha Zodiac's central feature – the sun god Helios driving his sun-chariot. Was the overall point to date the "Phaethon event" in their zodiac?

Myth and Astronomy

World mythology, and especially that of the Maya, indicate that the pole shift catastrophe expected in our near future comes at a time when the winter solstice sun is near the galactic center, where the arrow of Sagittarius and the stinger of Scorpio both point at it. But almost 13,000 years ago – half a precessional cycle back – the winter solstice sun was near the galactic anti-center, directly opposite the galactic center position between Sagittarius and Scorpio – between the horn tips of Taurus, the bull. Of course, this point of view is for lands in the northern hemisphere. When the Berezovka mammoth suddenly froze the vegetation inside it indicates it was summertime. But this was, after all, right before a pole shift. We have been assuming it was probably a thirty degree pole shift. What if we are wrong about that, and pole shifts overshoot the equator? It could have been a 150 degree pole shift, and the mammoths could have changed hemispheres. We must realize that the solstice sun near the galactic center is the winter solstice sun for one hemisphere, while at the same time it is the summer solstice sun for the other hemisphere.

And it does not require 150 degree movement during a pole shift to make people pay attention to both solstices. Multiple traditions from cultures on opposite sides of the equator would warn about a pole shift happening when the solstice sun was at the galactic center, but southern and northern hemisphere cultures would have different answers for whether it was the summer or winter solstice. Atlantis in Antarctica, or Inca ancestors in South America, might have an answer opposite to the Egyptian or Greek opinion of what season it was. If our ancient ancestral astronomer-priests had maintained enough knowledge of such cataclysms from multiple events and multiple locations that they were trained to watch both the summer solstice sun and the winter solstice sun, they might have also paid attention when the winter solstice sun was in the horns of Taurus.

Either option could explain why the Sampo was said to be broken with the help of an enormous bull. It may also be why Hathor, Egypt's goddess of destruction, is depicted wearing a "hat" made of the sun disk in the horns of a bull – if the last perceived episode of her wrath against humanity occurred during a pole shift almost 13,000 years ago and they were watching for both the summer solstice sun near Scorpio and the winter solstice sun at Taurus. Even the Babylonian god Marduk (so closely associated with Nibiru) has a name which was probably pronounced like "Marutuk" and could be derived from "amar" (bull calf) and "Utu" (the sun god) which could easily mean – the sun in Taurus.

The gap between Sagittarius and Scorpio is the region of sky the Maya were concerned about in their myths and calendar, warning us that when the winter solstice sun reaches the galactic center and the ball game (in many ways representing the world age) ends. The Maya also refer to this spot as "a black hole," "the Black Road, through the cleft in the Milky Way," "the hole in the middle of the four roads," and as "the Court of Creation in the Land of Death."[317] It is even the region of sky in which Christ overcomes Death, for the Valley of the Shadow of Death is the space between Sagittarius and the stinger of Scorpio at the galactic center.

I am reminded of Christianity's take on destruction at the time of this astronomical alignment in the Book of Revelation. Revelation Chapter 2 opens by describing God as "The One who holds the seven stars in His right hand, the One who walks among the seven golden lampstands." Could these seven be the seven "planets" of ancient times, the seven "planets" of alchemy – the seven lights in the sky that wander amongst the other fixed stars – the sun, the moon, Mercury, Venus, Mars, Jupiter, and Saturn? Revelation 2:5 warns us "I am coming to you and will remove your lampstand out of its place." (God's active force is coming to remove our planet's position?) Revelation 3:3 warns "If you do not wake up, I will come like a thief." When a temple priest had a shift overnight in the temple to protect the valuables and keep the fire going, the high priest would make his rounds to make sure the other priest had not fallen asleep on his watch. If he was awake, it was a meeting of friends at an expected time. But if the priest on overnight duty fell asleep in the temple, the high priest would wake him by setting his robe on fire, surprising him like a thief.

We should understand this analogy means that God comes on schedule, at an expected, predictable time, to those who are "awake" to the knowledge of the cycles... Revelation 3:13 "He who has an ear, let him hear" – those who are initiated, understand. Revelation 3:21 "I will grant to him to sit down with me on my throne, as I also overcame and sat down with my father on His throne." Wow, that sounds reminiscent of cycles of Egyptian "gods" in astronomical alignments in cycles... Revelation 4:1-2 "behold, a door standing open in heaven... and behold, a throne was standing in heaven, and One sitting on the throne." Revelation 4:6-7 "And there were seven lamps of fire burning before the throne [the seven wandering "planets"?]... and before the throne there was something like a sea of glass, like crystal; [the Milky Way?] and in the center and around the throne, four living creatures full of eyes in front and behind. The first creature was like a lion, and the second creature like a calf, and the third creature had a face like that of a man, and the fourth creature was like a flying eagle."

These are clearly the four cardinal constellations of the ancient zodiac, specifically Leo the Lion, Taurus the Bull, Aquarius the Man, and the last is less clear to modern readers only because Scorpio is not only represented as a scorpion but also as an eagle and as a serpent. The throne, described as being in the zodiac, must be within the band of the ecliptic. If the sea of glass is the Milky Way, then it must sit where the two intersect, at the galactic center. Revelation 5:1-7 describes a book of seven seals. Imagine a book with a cylinder lock where 7 circles must be dialed to the right position to open and unlock. Perhaps when the seven traditional "planets" are in the right alignment, the book opens...

This book will be "in the right hand of Him who sat on the throne" and Revelation continues to describe that the Lamb "came and took the book." This seems to describe the (winter solstice) sun coming to the galactic center over time. In John 20:17 Jesus says: "I am not yet ascended to my Father." In Revelation 5:1 Christ has reached the right hand of God, sits with him at the throne, and is worthy to open the book with seven seals that unleashes great destruction on the earth. At the point of alignment, the Lamb starts to break the seven seals of destruction. Revelation 6:12-14 includes details like "a great earthquake; and the sun became black as sackcloth... and the stars of the sky fell to the earth... The sky was split apart like a scroll when it is rolled up, and every mountain and island were moved out of their places." That's one way to describe a pole shift.

Revelation chapter 12 seems to describe some things over again from a new perspective. The dragon's tail "swept away a third of the stars of heaven and threw them to the earth." The chapter begins with "A great sign appeared in heaven" in which Virgo the Virgin gives birth to a male child, the dragon is poised to devour the child, "and her child was caught up to God and His throne." "And there was war in heaven, Michael and his angels waging war with the dragon. The dragon and his angels waged war, and they were not strong enough, and there was no longer a place found for them in heaven.

And the great dragon was thrown down." This sounds like every mythical battle of the gods, in which one regime overthrows another pantheon of "gods" and the old pantheon is thrown down to Tartarus like stars falling below the horizon in a pole shift. In this particular case I assume it is Draco, the dragon constellation at the North Celestial Pole, which "falls down" or is "thrown down" from his polar throne. A pole shift moving the earth's crust any substantial number of degrees of latitude will appear to cast the appropriate fraction of fallen stars/angels below the horizon of observers on the ground.

The Book of Enoch did not make the cut in our modern Bible, but in chapter 18 an angel suggests to Enoch that stars are angels and described the punishment of disobedient stars: "These stars which roll around over the fire are those who, at rising time, overstepped the orders of God: they did not rise at their appointed time. And He was wroth with them, and He bound them for 10,000 years until the time when their sin shall be fulfilled." This approximates the 12,960 year half-precession cycle that seems to pass between pole shifts – when the position of the stars will appear to change again. The constellation Draco, currently hovering over our North Pole, will no longer seem to hold the high position in the sky that other stars seem to circle around after a pole shift, the axis will point elsewhere, and the dragon will have fallen.

By Revelation Chapter 21 we get "a new heaven and a new earth, for the first heaven and the first earth passed away." This is another way to describe the post pole shift conditions in which the world is vastly different and the view of the sky, as seen by survivors, has shifted significantly. "He who sits on the throne" says "I am the Alpha and the Omega, the beginning and the end." He rules over everything, including the cycles of catastrophes that end and begin world ages. "The tabernacle of God is among men, and He will dwell among them." Does this mean that the galactic center is now visible as a bright blue light or Eye in the sky? That God is visibly present as He was when Adam and Eve were kicked out of Eden in a previous pole shift? Astrophysicist Carl Seyfert, who discovered that all spiral galaxies have periodically active galactic nuclei, said that the visible phase should last about a thousand years. Does this correspond to the Christian "millennium" after which things change and Satan must be released again after a thousand years?

The Hopi natives of North America describe the axial poles as twins that hold the world in place. At the end of every world age, the twins are instructed to leave their posts and let the world plunge into the chaos of Nuutungk Talonguaqa – the last day. This last occurred with a great shuddering and splintering of ice, when Sotuknang came to destroy the earth. Fortunately, many were warned to go underground with the Ant-people, where some survived until the earth started rotating normally again. "The twins had hardly abandoned their stations when the world, with no one to control it, teetered off balance, spun around crazily, then rolled over twice. Mountains plunged into the sea with a great splash, seas and lakes sloshed over the land... Waves higher than mountains rolled in upon the land. Continents broke asunder and sank beneath the seas."[318]

Native Americans in the northwestern United States have legends about their ancestors surviving a flood on various tall mountains in the region. Mt. Shasta, Mt. Jefferson, Mt. Baker, Mt. Rainier are all mentioned in stories of a great flood that rose almost to the tops of the mountains. The Yurok tribe in Northern California tell a story of how Thunder and Earthquake brought the ocean where it is today. "First Kingfisher and Earthquake went to the north end of the world.... Then Earthquake ran around. He ran around and the ground sank. It sank there at the north end of the world.... 'Now we will go to the south end of the world,' said Earthquake.... Thunder... will help us... the water will extend all the way to the south end of the world... Together the three of them went north... The Earth quaked and quaked and water flowed over it... Kingfisher emptied his shell and it filled the ocean halfway to the north end of the world... Now Thunder and Kingfisher and Earthquake looked at the ocean. 'This is enough,' they said.... So it is that the prairie became ocean."[319]

Natives of the Lake Tahoe region say: "the Great Spirit sent an immense wave across the continent from the sea."[320] The Spokane Indians describe a giant lake which used to cover the region. (We now know that a giant lake, Lake Missoula, did once cover the area, until the end of the last ice age.) The legend says that one summer morning the earth started shaking, and waves like mountains of water rose over the land. Then the sun was blotted out, and for days the earth kept rumbling and quaking. A rain of ashes fell for several weeks. Then a new river formed (the Spokane River) and drained Lake Missoula into a big river to the west (the Columbia River.) The nearby Cowichan Indians in British Columbia say that "The people who populated the earth long ago had wise men [astronomers?] who could foretell the future..." They "dreamed" of destructive floods and set to work to build a ship. "...When the rain began, the wise men and the people who had believed their warnings took their families and food on board" and sailed to safety on a mountaintop.[321]

In The Bear Tribe's Self-Reliance Book, we are told a similar version of Hopi myths of the Second World, in which "the people lived until they forgot their origin and grew cold and hard to the ways of the Good Life. And so, once again, Sotuknang was ordered to destroy the world. This time he ordered the Twins, Poquanghoya and Palongawhoya, to leave their stations at the North and South Poles and let the world be destroyed. They did this, after the people had again hidden with the Ant People underground. After the Twins left their stations, the world's stability was removed and so it flipped end over end and everything on it was destroyed by ice." Other North American tribes also seem to indicate a pole shift when they link stories of the last day with the North Pole. The Skidi Pawnee say "When the time comes for all things to end..." "The command for the ending of all things will be given by the North Star, and the South Star will carry out the commands."[322]

Hopi legends say that in the third world, people had great technology like flying aircraft, but they turned evil and were destroyed by waves taller than mountains. Chief Dan Katchongva was a leader of the Hopi Sun Clan, who said the Hopi were the survivors from a previous world who claimed all the land for the Great Spirit "as commanded by Massau'u and for the True White Brother who will bring on Purification Day."[323] In the legend, the Great Spirit made two stone tablets to contain all the instructions, knowledge, and prophetic warnings needed in the Fourth World, and were given to the Hopi chief to take to a new land. His sons each took one of the stone tablets and set off east and west, to return after teaching the world how to rout wickedness from the land. The two brothers would recognize each other from the tablets each held, and also because the older brother would have white skin and a red cloak. Some clues indicate the older brother went to Tibet.

Prophecies of Pole Shifts

Before the Hopi brothers are reunited to bring on the day of purification there are many earlier signs that come first. Allegedly:

White men will bring wagons tied together pulled by something besides horses – trains?
There will be cobwebs in the air and lines like snakes across the land – telephone lines, highways, railroads, jet vapor trails...
Man will invent a weapon called "the gourd of ashes" which can burn the land and boil rivers – atomic bombs?
Two brothers will build a ladder to the moon – 1969 moon landings?
White men will build houses in the sky – space stations?
People of the world will gather in "a house of mica" on the east coast of North America – the glass exterior of the United Nations Building in New York?

If these signs from Hopi legends are meaningful, then it would seem the prophetic prerequisites for the end of the present world age have come to pass, and that just as so much other evidence indicates – the next pole shift could happen soon. In Hopi mythology Sotuknang created nine worlds; "the first three of these worlds, Tokpela, Tokpa and Kuskurza have already been inhabited and subsequently destroyed due to the corruption and wickedness of man. Each time one of the worlds is destroyed, the faithful Hopi are taken underground and saved from destruction to later emerge and populate the next world. According to [Frank] Waters' books... Hopis believe that humanity is currently residing in the fourth world, Túwaqachi. Like the previous worlds, Túwaqachi is also prophesied to be destroyed because of the corruption of humanity."[324] The final sign, some claim, is the appearance of a blue star. "In Hopi mythology, the Blue Star Kachina or Saquasohuh, is a kachina or spirit, that will signify the coming of the beginning of the new world by appearing in the form of a blue star."[325]

In my own books, I have suggested that the next pole shift could be due in the early 21st century. Bible prophecies describe heavenly bodies in the night sky during the end times visions of Daniel, Isaiah, and John. If we compare these descriptions with astronomical software that tells us where all heavenly bodies are at any given time, we can see a correlation to the week of December 21-28, 2019. Not only does this match a staggering number of other, nonastronomical clues in Bible prophecy – it also spans the duration of Hanukkah – which is very symbolic of destruction, cleansing, purification, and re-dedication, and the fulfillment of end times prophecy. Most significant to me, this week also sees the sun, moon, and planets move in ways that act out all major steps of an ancient, week-long Jewish wedding ceremony. Macrobius tells us that in Egypt's Osirian legends, Osiris is the sun and the husband, and Isis in the earth and the bride. We have noted the astronomical home of Vishnu, and India's legends tell us the goddess earth is Vishnu's bride... Should we assume this celestial wedding – a kind of stellar mystery play – also symbolizes Jesus coming for His bride as the pole shift makes a new heaven and a new earth? I cover this correlation in detail in End Times and 2019.

The Maya made sure our attention is focused on this time frame as well. They were absolutely obsessed with the astronomical alignment of the winter solstice sun and the galactic center on some future 21st of December. Because the sun's disk appears to be about one half of a degree wide, the alignment of the solstice sun and the galactic axis is about one half of one 360th of a precession cycle – about 36-38 years. John Major Jenkins said the alignment is in effect from about 1980-2018, so December 21, 2019 is the first time the winter solstice sun is past the galactic equator. In Mayan thinking, this is the astronomical equivalent of the end of their ballcourt game, when the ball (the sun) finally passes through the ball hoop – game over.

The end of the Mayan Long Count calendar on December 21, 2012 may have been meant to initiate a final seven year period like the Christian tribulation, culminating with the end of the present world age via a cosmic flood, as depicted in the Dresden Codex of the Mayan Popul Vuh. The astronomical alignment detailed in the flood drawing is not in effect in 2012, but is accurate for December 2019. Their monuments were also built with the future in mind; at Chichen Itza the pyramid synchronizes with the astronomical alignment it symbolizes when precession brings us to... the 21st century. John Major Jenkins wrote in Maya Cosmogenesis that the Pyramid of Kukulkan "is a precessional alarm clock set for the twenty-first century." As Graham Hancock commented years after 2012: "the story of the Mayan calendar isn't quite over yet."

Even the French psychic Nostradamus offered multiple prophecies of a pole shift, though he wrote nothing to suggest it would occur before WWIII. You have undoubtedly seen outrageous claims about his prophecies on the internet, with predictions for every new year coming out on Youtube in late December. Most such interpretations are utter nonsense. But Nostradamus does warn that "Libra abandons her Phaethon" (Libra is the balance, and Phaethon was Apollo's son, who couldn't handle his father's sun-chariot when he took the reins for a day. Phaethon losing his balance and having the sun veer through the sky would be one sign of a pole shift happening. In another warning Nostradamus wrote that at the end of time "the great translation will be made and it will be such that one will think the gravity of the earth has lost its natural movement…" Nostradamus also suggests that a pole shift will come at the end of WWIII when he warns: "After great trouble for humanity, a Greater one is prepared; the Great Mover renews the ages…" It reminds me that Bible prophecy's Judgment Day only comes after years of Great Tribulation, economic collapse and starvation, persecution by the Antichrist, and an apocalyptic world war. John Michell points out "that event, according to all prophecies, will take place at a period of extreme need and desperation."[326]

Gopi Krishna expresses a similar expectation – a pole shift right at the end of WWIII – much like Nostradamus. But he believes it will happen then as an act of divine intervention to limit the human suffering of nuclear war. Like many New Age psychics, Gopi Krishna believes that pole shifts come from imbalances caused by man. Not from technology destroying the Earth but from evil behavior. This idea is global and ancient, from Navajo stories that "this Fourth World was destined to be destroyed when the people do not live the right way"[327] to biblical stories of Adam and Eve breaking God's laws in "Eden" or when "the Lord saw that the wickedness of man was great" in the days of Noah. (Genesis chapter 6) As the New Age Reverend Paul Solomon wrote: "there never would have been the shifting of the poles upon this planet if there were not the creating of conditions among men that were defiant to the laws of God… the shifting of the poles was caused by the activity of men in defiance of their God."[328]

Gopi Krishna, in books like <u>The Shape of Events to Come</u>, suggests (as paraphrased by John White in <u>Pole Shift</u>, p. 269) "Imagine the human race struggling in the midst of a nuclear war. And then imagine… that the very ground beneath the armies begins to lurch and crumble. Whether such an event is interpreted as supernatural intervention or the outcome of some natural process touched off by atomic explosions, the result could be only one thing: the cessation of hostilities… Such a development, from Gopi Krishna's perspective, would be divine intervention through natural means" to end WWIII and limit humanity's destruction of itself and the world.

After studying Nostradamus prophecies for several decades, I conclude that Nostradamus expects a pole shift at the end of a Third World War between Islam and the West. He repeatedly describes a long conflict between Islamic nations and the Christian West, gradually intensifying over 27 years and culminating in WWIII and immediately followed by a pole shift. In the 1990s, I originally thought this could describe warfare during the years from 1991-2018, with a pole shift the next year. But it is already 2018 as I edit the book you are reading now, leaving very little time for WWIII to precede a pole shift in 2019… At this point It seems much more likely that Nostradamus' 27 years of increasing hostilities between Islam and the West start with September 11, 2001 – and end with a pole shift around 2028-2029. I cover these prophecies, and many more describing WWIII, in <u>Nostradamus and the Islamic Invasion of Europe</u>.

So when is the next pole shift due? That depends on which sources of information we choose to believe. 2019? 2029? Can we even agree on a broad range of at least a few decades in the early to mid-21st century with any certainty?

Surviving the Pole Shift

How should you prepare for surviving the upcoming pole shift? There are many books and commentaries with specific advice on this topic, from constructing underground bunkers to relocating to the Colorado Plateau or New Zealand. It only takes a few seconds to search key words and come up with book titles like: Surviving 10: Surviving the Pole Shift or Surviving After the Pole Shift. But is it really worth worrying about? Other titles include: Crossing the Cusp: Surviving the Edgar Cayce Pole Shift, Planet X Forecast and 2012 Survival Guide, and How To Survive 2012.

The "Edgar Cayce Pole Shift" didn't happen when his followers expected it in 1998. There was no pole shift when many people expected it as the Mayan Long Count Calendar came to an end in December 2012. My astrological interpretation of biblical prophecy points to a pole shift the week of December 21-28, 2019. The biblical prophet Ezra repeatedly asked about the timing of the end, and was told in 2 Esdras 14:11-12 "the age is divided into twelve parts, and nine of its parts have already passed, as well as half of the tenth part; so two of its parts remain, besides half of the tenth part." I interpret that to mean that each part is a thousand years, and that approximately 2&1/2 thousand years after Ezra wrote that around 500 B.C. we can expect the end of the present age via another pole shift, as he described the last one in 2 Esdras 3:9-18 "But again, in its time you brought the flood upon the inhabitants of the world and destroyed them.... You bent down the heavens and shook the earth, and moved the world, and caused its depths to tremble.... Your glory passed through the four gates of fire and earthquake and wind and ice."

Nostradamus seems to suggest a pole shift just over 27 years after a state of war between Islam and the West begins – which could be considered to have started in 1991, or 2001. (Delores Cannon claimed Nostradamus says the pole shift will be in October 2029.) The Maya gave us warnings alerting us to 2012, 2019, and the 21st century overall. If forced to offer my best guess on exact timing, I would still argue that many signs point to late December 2019, but I realize that such a timeline for a pole shift may have already come and gone uneventfully by the time you have read this. Despite all the evidence, clues, and coincidences I have seen that convince me there is a cyclic catastrophe that is due again "soon" – I can't guarantee that it happens in on any particular day. If you want to be prepared for it, don't focus on dates. Focus on choosing a location and on having a plan.

Then there is the question of how to prepare. Zetatalk's Nancy Leider has said that people should plan to be at least a hundred miles inland from coastlines and at elevations at least two hundred feet above sea level to avoid the associated tsunamis. I think this suggestion, allegedly coming as a warning from aliens, would be woefully inadequate for surviving a pole shift. At the other extreme, Chan Thomas suggests that tsunamis over two miles high (well over 10,000 feet) will sweep over entire continents, overcoming any survival preparations and killing off almost everyone. We don't really know for certain how many degrees of latitude a future pole shift will cover, or whether it will be done in an hour, or a day, or a week, or if it takes far longer. We don't know with any certainty what latitude any particular location will end up at after the event, or whether it will remain above sea level or not. There is no guarantee that your location, or any particular location – will be safe.

There are reasonable considerations that will improve your chances, including moving to higher ground and away from coastlines. I wouldn't want to be near an earthquake fault line, an active volcano, a nuclear reactor, a chemical plant, or a city full of starving, desperate survivors. Ideally I would want to survive a pole shift far from all of these things, in a squat, sturdy house on a farm in a rural area with many sources of fresh water. I would want to have many supplies that cannot be reproduced without modern manufacturing and advanced technology. I would want to already be a part of a community of survivalists, or at least one of people who have the skills needed to farm, hunt, build and fix things, and defend us from desperate hordes.

But most of us have a job we need to pay the bills. We often live near family and friends. How many of us have the financial resources, or the mental and emotional fortitude, to pack up and relocate to a more remote survival location because we think there might be a pole shift coming? Very few of us could easily achieve that even if we thought we should. Granted, some of you may be in a situation where you have to find a new job and new home soon anyway. If I were, for example, a recently unemployed breadwinner with highly desired skills and had my pick of new jobs, I might lean towards rural Colorado over New York City in my job searches.

My advice for most people, however, is not to worry too much about it. It doesn't seem worth turning your life upside down on the chance that such events are coming soon and that by devoting your life to surviving it you would make all the right guesses and come out on top. Without a golden ticket onto some kind of submarine ark or secret government base built deep into the Earth or the moon... Our odds of surviving through a catastrophic pole shift while living a normal life on the surface, even when aware of all the information we do have, remain slim. It will largely come down to specific details we cannot yet predict with certainty: Where will the new North and South Poles be located? How high will tsunamis rise? When, exactly, will such an event occur?

At the moment, we can only "guesstimate" such crucial details. But there is a chance that dramatic events will start to occur slowly enough that you will have time to make a move. If a bright new blue "star" appears at the galactic center between Sagittarius and Scorpio; or if the sun does not rise at the appropriate time or location; if there is a sudden change in sea level of even a few feet; if television and the internet are buzzing with the news that the poles have unquestionably started to shift, do not plan to wait a few days for confirmation – assume events will accelerate quickly! Be ready to pack up the most important people and supplies and head to that rural location you have in mind immediately!

A condensed version of Matthew 24:16-42 seems appropriate for a day when you might observe such events starting: "flee to the mountains... [do] not go down to get the things out of the house... for then there will be a great tribulation, such as has not occurred since the beginning of the world until now... the stars will fall from the sky, and the powers of the heavens will be shaken. And then the sign of the Son of Man will appear in the sky... Heaven and earth will pass away... the coming of the Son of Man will be just like the days of Noah... they did not understand until the flood came and took them all away... Therefore be on the alert, for you do not know which day..."

Selected Bibliography

Brown, Hugh A. Cataclysms of the Earth. New York, NY: Twayne Publishers, 1967

De Santillana, Giorgio & Hertha von Dechend. Hamlet's Mill: An Essay Investigating the Origins of Human Knowledge and its Transmission Through Myth. Jaffrey, NH: David R. Godine, 1969.

Flem-Ath, Rand & Rose Flem-Ath. When the Sky Fell: In Search of Atlantis. New York, NY: St. Martin's Press, 1995

Hapgood, Charles. Earth's Shifting Crust : A Scientific Key to Many of Earth's Mysteries. New York, NY: Pantheon Books, 1958

Hapgood, Charles. Maps of the Ancient Sea Kings: Evidence of Advanced Civilization in the Ice Age. Kempton, IL: Adventures Unlimited Press, 1966

Hapgood, Charles. The Path of the Pole. Philadelphia, PA: Chilton Books, 1970

LaViolette, Paul. Earth Under Fire: Humanity's Survival of the Ice Age. Rochester, VT: Bear & Company, 1997

Lemesurier, Peter. The Great Pyramid Decoded. Rockport, MA: Element Books, 1996

Jenkins, John M. Maya Cosmogenesis 2012. Santa Fe, NM: Bear & Company, 1998

Michell, John. The Dimensions of Paradise: Sacred Geometry, Ancient Science, and the Heavenly Order on Earth. Rochester, VT: Inner Traditions, 2008

Montaigne, David. End Times and 2019. Kempton, IL: Adventures Unlimited Press, 2013

Thomas, Chan. The Adam and Eve Story. Los Angeles, CA: Emerson House, 1965

White, John. Pole Shift: Scientific Predictions and Prophecies About the Ultimate Natural Disaster. Virginia Beach, VA: A.R.E. Press, 1980

End Notes

[1] Albert Einstein wrote the foreword to Charles Hapgood's Earth's Shifting Crust : A Scientific Key to Many of Earth's Mysteries. New York, NY: Pantheon Books, 1958

[2] Newton, Isaac. Principia Mathematica. Section 1, Prop. 66, Theory 26, Cor. 22 London, UK: 1687

[3] http://custance.org/Library/Volume6/Part_III/Chapter2.html

[4] Hancock, Graham & Robert Bauval. The Message of the Sphinx. New York, NY: Three Rivers Press, 1996, p. 201, citing E. Reymond's Mythical Origin of the Egyptian Temple, pp. 106-107

[5] Maud Makemson, Maud. The Morningstar Rises, Yale U. Press, 1941, p. 60

[6] Huxley, Thomas. Quote cited in Hapgood's Path of the Pole, p. 294, and also in Hancock's Fingerprints of the Gods, p. 460 – apparently from The Quarterly Journal of the Geological Society of London, vol. 25, 1869, p. xlvi

[7] Horberg, Leland. "Radiocarbon Dates and Pleistocene Geological Problems of the Mississippi Valley Region." Journal of Geology, v. 63, May 1955, p. 285)

[8] https://news.nationalgeographic.com/2018/01/earth-magnetic-field-flip-north-south-poles-science/

[9] http://www.poleshiftnews.com/global-leaders-view-point.html

[10] 2/27/16, yournewswire.com/nasa-warn-that-earth-is-soon-going-to-experience-a-full-pole-shift/

[11] Nationalgeographic.com, 1/31/2018

[12] http://www.poleshiftnews.com/how-serious-is-a-pole-shift.html

[13] Opdyke, Neil. "Paleomagnetism, polar wandering, and the rejuvenation of crustal mobility." Journal of Geophysical Research, vol. 100, Issue B12, December 1995

[14] Hapgood, Charles. Earth's Shifting Crust: A Scientific Key to Many of Earth's Mysteries. New York, NY: Pantheon Books, 1958, p. 184

[15] https://globalrumblings.blogspot.com/2012/11/pole-shift

[16] https://www.usgs.gov/faqs/it-true-earths-magnetic-field-occasionally-reverses-its-polarity

[17] "Magnetic Pole Reversal Ahead?" 1/10/2017, earthsky.org/earth/magnetic-pole-reversal-ahead

[18] Hall, Jessica. "South Africa may be the epicenter of a geomagnetic pole reversal in progress." Extreme Tech, 2/10/2017

[19] https://en.wikipedia.org/wiki/World_Magnetic_Model

[20] https://www.ngdc.noaa.gov/geomag/WMM/DoDWMM.shtml

[21] "Scientists Solve Mystery of Earth's Shifting Poles" 11/9/2012 https://www.npr.org/2012/11/09/164797147/scientists-solve-mystery-of-earths-shifting-poles

[22] Sweeney, Emmet. "Atlantis: The Evidence of Science." 2010 (page numbers not shown in pdf online)

[23] LaViolette, Paul. Earth Under Fire: Humanity's Survival of the Ice Age. Rochester, VT: Bear & Company, 1997, p. 3

24 Jenkins, John M. Maya Cosmogenesis 2012. Santa Fe, NM: Bear & Company, 1998, p xxxiii

25 Kak, Subhash. "Birth and Early Development of Indian Astronomy." in Astronomy Across Cultures: The History of Non-Western Astronomy. Dordrecht, Netherlands: Kluwer, 2000, p. 311

26 Michell, John. The Dimensions of Paradise: Sacred Geometry, Ancient Science, and the Heavenly Order on Earth. Rochester, VT: Inner Traditions, 2008, p. 55

27 LaViolette, Paul. Earth Under Fire: Humanity's Survival of the Ice Age. Rochester, VT: Bear & Company, 1997, p.26

28 LaViolette, Earth Under Fire, p. 308

29 LaViolette, Earth Under Fire, p.2

30 Michell, John. The Dimensions of Paradise: Sacred Geometry, Ancient Science, and the Heavenly Order on Earth. Rochester, VT: Inner Traditions, 2008, p. 137

31 Hapgood, Charles. Earth's Shifting Crust : A Scientific Key to Many of Earth's Mysteries. New York, NY: Pantheon Books, 1958, p. 245

32 Ma, Ting-Ying, Research on Past Climate and Continental Drift (v. 6): The Sudden Total Displacement of the Outer Solid Earth Shell by Slidings Relative to the Fixed Rotating Core of the Earth, 1953, p. 13 and p. 5

33 Hapgood, Charles. Earth's Shifting Crust : A Scientific Key to Many of Earth's Mysteries. New York, NY: Pantheon Books, 1958, p. 71-76

34 Brown, Hugh A. Cataclysms of the Earth. New York, NY: Twayne Publishers, 1967, p. 69

35 Graham, John W. "Evidence of polar shift since Triassic time." Journal of Geophysical Research, Vol. 60, Issue 3, September 1955.
https://agupubs.onlinelibrary.wiley.com/doi/pdf/10.1029/JZ060i003p00329

36 Hapgood, Charles. The Path of the Pole. Philadelphia, PA: Chilton Books, 1970, pp. 14-16

37 LaViolette, Paul. Earth Under Fire: Humanity's Survival of the Ice Age. Rochester, VT: Bear & Company, 1997, p. 202)

38 Coe, Robert & Michel Prevot. "Evidence Suggesting Extremely Rapid Field Variation During a Geomagnetic Reversal." Earth and Planetary Science Letters Volume 92, Issues 3–4, April 1989,

39 Witze, Alexandra."Earth's Magnetic Field Flipped Superfast," Wired.com, 9/2/2010

40 Farrand, William. "Frozen Mammoths and Modern Geology." Science Magazine, March 17, 1961

41 Sanderson, Ivan. "Riddle of the Frozen Giants." Saturday Evening Post, January 16, 1960

42 Tolmachev, I.P. "The Carcasses of the Mammoths and Rhinoceroses Found in the Frozen Ground of Siberia." American Philosophical Society, Philadelphia, 1929, pp. 51, 57

43 Sputnik Magazine, November 1968, p. 54

44 Hapgood, Charles. The Path of the Pole. Philadelphia, PA: Chilton Books, 1970, p. 251

[45] Bennett, Kyle. "Mammoths of the Last Polar Age." 2011, http://pathofthepole.yolasite.com/arktos---a-polar-myth.php
[46] Bennett, Kyle. "Mammoths of the Last Polar Age."
[47] Brown, Hugh A. Cataclysms of the Earth. New York, NY: Twayne Publishers, 1967, p 28
[48] Hibben, Frank. The Lost Americans. New York, NY: T.Y. Crowell, 1968, pp. 90-92
[49] Hibben, Frank. The Lost Americans, pp. 168-169
[50] Hibben, Frank. The Lost Americans, pp. 168-178
[51] Hapgood, Charles. The Path of the Pole. Philadelphia, PA: Chilton Books, 1970, p. 182
[52] "Discussion with Robert Felix / Magnetic Reversal Ice Age" 3/9/2018 www.youtube.com/watch?v=W_uiNwoSZu4
[53] Hapgood, Charles. The Path of the Pole. Philadelphia, PA: Chilton Books, 1970, p. 46, citing p. 246 of A.P. Coleman's Ice Ages Recent and Ancient, 1929
[54] Flem-Ath, Rand & Rose Flem-Ath. When the Sky Fell: In Search of Atlantis. New York, NY: St. Martin's Press, 1995, p. 80
[55] The London Times, December 5, 1994 – as cited in Hancock's Fingerprints of the Gods, p. 478
[56] Hapgood, Charles. The Path of the Pole. Philadelphia, PA: Chilton Books, 1970, citing A.P. Coleman's Ice Ages Recent and Ancient, 1929, pp. 7-9
[57] Hapgood, Charles. The Path of the Pole. Philadelphia, PA: Chilton Books, 1970, citing A.P. Coleman's Ice Ages Recent and Ancient, 1929, p. 262
[58] Hancock, Graham. Fingerprints of the Gods: The Evidence of Earth's Lost Civilization. New York, NY: Three Rivers Press, 1995, p. 461
[59] Hapgood, The Path of the Pole, p. 144
[60] Hapgood, The Path of the Pole, p. 203
[61] Personal letters between Einstein and Hapgood, cited on p. 223 of The Path of the Pole
[62] Symposium on the Crust of the Earth, Geological Society of America, 1955, p. 320
[63] Pirsson, Louis and Charles Schuchert, A Textbook of Geology, New York, NY: John Wiley, 1929, p. 404
[64] Suess, Eduard. Das Antlitz Der Erde (The Face of the Earth), v. 1, 1904, pp. 17-18
[65] Dutton, Clarence. "On Some of the Greater Problems of Physical Geology, in v. 78 of Physics of the Earth, 1931, pp. 201-202
[66] Bucher, Walter. The Deformation of the Earth's Crust, Princeton, NJ: Princeton University Press, 1933, p. 144
[67] Umbgrove, J., The Pulse of the Earth, The Hague, Netherlands, 1947, p. 31
[68] Umbgrove, The Pulse of the Earth, p. 23
[69] Fowden, Garth. The Egyptian Hermes. New York, NY: Cambridge University Press, 1986, p. 33
[70] Hapgood, Charles. Maps of the Ancient Sea Kings: Evidence of Advanced Civilization in the Ice Age. Kempton, IL: Adventures Unlimited Press, 1966, p. 89
[71] Hapgood, Maps of the Ancient Sea Kings, p. 149

[72] Personal correspondence with Hapgood, cited in multiple books.
[73] Hancock, Graham. Fingerprints of the Gods: The Evidence of Earth's Lost Civilization. New York, NY: Three Rivers Press, 1995, pp. 88-91
[74] d'Orbigny, Alcide. Voyage dans l'Amerique Meridionale. Paris, 1842, v. 3, pp. 82-86
[75] Darwin, Charles. Voyage of the Beagle. London, UK: J. Murray, 1913, p. 178
[76] Flem-Ath, Rand & Rose Flem-Ath. When the Sky Fell: In Search of Atlantis. New York, NY: St. Martin's Press, 1995, p. 79 – citing The Paleoecology of Beringia, 1982, p. 309
[77] Hibben, Frank. The Lost Americans. New York, NY: T.Y. Crowell, 1968, p. 168
[78] Scott, W.B. A History of Land Mammals in the Western Hemisphere. New York, NY: MacMillan, 1937, p. 75
[79] Roche, D. et al. Duration and iceberg volume of Heinrich event 4" Nature: 432, p. 380, 2004
[80] http://slideplayer.com/slide/4314982/
[81] Brown, Hugh A. Cataclysms of the Earth. New York, NY: Twayne Publishers, 1967, p. 38
[82] LaViolette, Paul. Earth Under Fire: Humanity's Survival of the Ice Age. Rochester, VT: Bear & Company, 1997, p. 189
[83] Channell, J. et al. "Magnetic record of deglaciation using FORC-PCA, sortable-silt grain size, and magnetic excursion at 26 ka, from the Rockall Trough (NE Atlantic)" Geochemistry, Geophysics, Geosystems. Vol 17, issue 5, May 2016
[84] Van Flandern, Tom. Dark Matter, Missing Planets, and New Comets. Berkeley, CA: North Atlantic Books, 1994, p. 430
[85] Atkinson, Nancy. "Pole Shift on Europa." in Universe Today, May 14, 2008
[86] Wood, Anthony. "A massive asteroid impact could have caused Enceladus to tip onto its side" in New Atlas, May 31, 2017.
[87] Massey, Gerald. Ancient Egypt: The Light of the World. London, UK: T. Fisher Unwin, 1907,
[88] Hutton, William & Jonathan Eagle. Earth's Catastrophic Past and Future: A Scientific Analysis of Information Channeled by Edgar Cayce. Boca Raton, FL: Universal Publishers, 2004, p. 213
[89] cited as 12,190-12,570 years ago in 1970, see Hapgood, Path of the Pole, p. 140
[90] Jenkins, John M. Maya Cosmogenesis 2012. Santa Fe, NM: Bear & Company, 1998, p. xxvii
[91] Warshofsky, Fred. Doomsday: The Science of Catastrophe. New York: Readers' Digest Press, 1977, pp.78-79
[92] Scofield, Bruce. "What Really Happened in 3100 B.C." http://www.onereed.com/articles/3000bc1.html)
[93] www.convertalot.com/asteroid_impact_calculator.html
[94] Haliburton, Robert. New Materials for the History of Man, 1868, p. 13
[95] Smyth, Charles Piazzi. Life and Work at the Great Pyramid. Edinburgh, UK: Edmonston and Douglas, 1867 pp. 382, 390-391
[96] Hagar, Stansbury. "November Meteors in Maya and Mexican Tradition" in Popular

Astronomy, v. 39, #7, 1931 p. 399

[97] Schwaller de Lubicz. R. A. The Temple In Man: Sacred Architecture and The Perfect Man. Rochester, VT: Inner Traditions, 1981 p. 106

[98] LaViolette, Paul. Earth Under Fire: Humanity's Survival of the Ice Age. Rochester, VT: Bear & Company, 1997 p. 239)

[99] Michell, John. The Dimensions of Paradise: Sacred Geometry, Ancient Science, and the Heavenly Order on Earth. Rochester, VT: Inner Traditions, 2008, p. 9

[100] Michell, Dimensions of Paradise, p. 135

[101] De Santillana, Giorgio & Hertha von Dechend. Hamlet's Mill: An Essay Investigating the Origins of Human Knowledge and its Transmission Through Myth. Jaffrey, NH: David R. Godine, 1969, p.5

[102] De Santillana, Hamlet's Mill, p.3

[103] De Santillana, Hamlet's Mill, pp. 310-311

[104] Michell, John. The Dimensions of Paradise: Sacred Geometry, Ancient Science, and the Heavenly Order on Earth. Rochester, VT: Inner Traditions, 2008 p. 121

[105] Michell, Dimensions of Paradise, p. 81

[106] Godwin, Joscelyn. Atlantis and the Cycles of Time. Rochester, VT: Inner Traditions, 2011, pp. 343-344

[107] Barton, George. "On the Babylonian Origin of Plato's Nuptial Number" Journal of the American Oriental Society, v 29 1908, p. 213

[108] Michell, Dimensions of Paradise, p. 86

[109] Lemesurier, Peter. The Great Pyramid Decoded. Rockport, MA: Element Books, 1996, p. 36

[110] Pageau, Jonathan. "Heaven Is Round. Earth Is Square." 11/13/2014 https://www.orthodoxartsjournal.org/heaven-round-earth-square/

[111] Fletcher, Rachel. "Squaring the Circle: Marriage of Heaven and Earth." Nexus Network Journal, vol. 9, #1, 2007 - https://link.springer.com/content/pdf/10.1007/s00004-006-0033-7.pdf

[112] Sora, Steven. The Triumph of the Sea Gods. New York, NY: Simon & Schuster, 2007

[113] Michell, John. The Dimensions of Paradise: Sacred Geometry, Ancient Science, and the Heavenly Order on Earth. Rochester, VT: Inner Traditions, 2008 p. 137

[114] Hancock, Graham. Underworld: The Mysterious Origins of Civilization. New York, NY: Three Rivers Press, 2003 p. 74

[115] Palmer, William. Egyptian Chronicles: With a Harmony of Sacred and Egyptian Chronology. London, UK: 1861, p.625

[116] Hapgood, Charles. The Path of the Pole. Philadelphia, PA: Chilton Books, 1970, p. 173

[117] Hapgood, The Path of the Pole, p. 103

[118] Horberg, Leland. "Radiocarbon Dates and Pleistocene Geological Problems of the Mississippi Valley Region." Journal of Geology, v. 63, May 1955, p. 286

[119] Weidner, Jay & Vincent Bridges. The Mysteries of the Great Cross of Hendaye: Alchemy and the End of Time. Rochester, VT: Inner Traditions, 2003 pp. 264-265

[120] Campbell, Joseph. The Mythic Image. Princeton, NJ: Princeton University Press,

1974 p. 149

[121] De Santillana, Giorgio & Hertha von Dechend. Hamlet's Mill: An Essay Investigating the Origins of Human Knowledge and its Transmission Through Myth. Jaffrey, NH: David R. Godine, 1969 p. 341

[122] Michell, John. The Dimensions of Paradise: Sacred Geometry, Ancient Science, and the Heavenly Order on Earth. Rochester, VT: Inner Traditions, 2008 pp. 62-63

[123] Montaigne, David. End Times and 2019. Kempton, IL: Adventures Unlimited Press, 2013 p. 113 – quoting Gavin Menzies' book 1434: The Year a Magnificent Chinese Fleet Sailed to Italy and Started the Renaissance, p.76 – itself quoting Fernando-Armesto's Columbus.

[124] Hancock, Graham & Robert Bauval. The Message of the Sphinx, New York, NY: Three Rivers Press, 1996, p. 242

[125] Newton, Isaac. Principia Mathematica. Section 1, Prop. 66, Theory 26, Cor. 22 London, UK: 1687

[126] Cuvier, Georges. "Essay on the Theory of the Earth." The Edinburgh Review. Vol. XXII, London, UK: Archibald Constable & Company, 1814, p. 458

[127] Darwin, Charles. Voyage of the Beagle. London, UK: J. Murray, 1913, p. 178

[128] Hapgood, Charles. The Path of the Pole. Philadelphia, PA: Chilton Books, 1970, pp. 306-307

[129] Meert, Joseph. "Widespread extinction after Earth's hyperactive magnetic field reversals." 2/23/2016, Watchers.com

[130] Simpson, George. The Major Features of Evolution. Columbia University, NY, 1953, p. 235

[131] Flem-Ath, Rand & Rose Flem-Ath. When the Sky Fell: In Search of Atlantis. New York, NY: St. Martin's Press, 1995, p. 43, citing Edward Lurie's Louis Agassiz, p. 98

[132] Evans, John. "On a Possible Geological Cause of Changes in the Position of the Axis of the Earth's Crust." Proceedings of the Royal Society of London, vol. 15, 1866-67

[133] Flem-Ath, When the Sky Fell, p. 67

[134] Blavatsky, Madame Helena. The Secret Doctrine. Vol. II, London, UK: Theosophical Publishing House p. 85

[135] Blavatsky, Madame Helena. The Secret Doctrine. Vol. IV, p. 103

[136] Blavatsky, Madame Helena. The Secret Doctrine. Vol. III, p. 152

[137] Blavatsky, Madame Helena. The Secret Doctrine. Vol. III, p. 329

[138] Blavatsky, Madame Helena. The Secret Doctrine. Vol. IV, p.294

[139] White, John. Pole Shift: Scientific Predictions and Prophecies About the Ultimate Natural Disaster. Virginia Beach, VA: A.R.E. Press, 1980, p. 314

[140] Childress, David Hatcher. Lost Cities and Ancient Mysteries of Africa and Arabia, 2002, p. 61 and from Diane Stein, in her book: Prophetic Visions of the Future, in which she is commenting on David Hatcher Childress' writing on the Lemurian Fellowship, which is based on channeled spiritual messages.

[141] Cayce, Edgar. Reading 3976-15, 1/19/1934

[142] Cayce, Edgar. Reading 826-8, 8/11/1936

[143] Cayce, Edgar. Reading 2746-2, 11/11/1943

[144] Cayce, Edgar. Reading 3976-15, 1/19/1934
[145] Cayce, Edgar. Reading 5249-1, 6/12/1944
[146] Brown, Hugh A. Cataclysms of the Earth. New York, NY: Twayne Publishers, 1967
[147] Brown, Cataclysms of the Earth. p. 145
[148] Brown, Cataclysms of the Earth, p. 139
[149] Frankfort, Henri. Ancient Egyptian Religion. New York, NY: Harper & Row, 1948, p. 13
[150] Brown, Cataclysms of the Earth, p. 118
[151] Brown, Cataclysms of the Earth, p. 8
[152] Brown, Cataclysms of the Earth, p. 102
[153] Bogue, Scott, et al. "Directional change during a Miocene R-N geomagnetic polarity reversal recorded by mafic lava flows, Sheep Creek Range, north central Nevada, USA." Geochemistry, Geophysics, Geosystems. Vol. 18, issue 9, September 2017
[154] Emory, Kenneth & Elazar Uchupi. The Geology of the Atlantic Ocean. Berlin, Germany: Springer Books, 1984, p. 859
[155] Brown, Cataclysms of the Earth, pp. 10, 117
[156] Brown, Cataclysms of the Earth, p. 123
[157] Brown, Cataclysms of the Earth, pp. 146, 273
[158] Pauley, K.A. "The Cause of the Great Ice Ages." Scientific Monthly, August 1952
[159] Velikovsky, Immanuel. Earth in Upheaval. New York, NY: Doubleday, 1955 p. 126 & 239
[160] White, John. Pole Shift: Scientific Predictions and Prophecies About the Ultimate Natural Disaster. Virginia Beach, VA: A.R.E. Press, 1980 p. 287
[161] Velikovsky, Immanuel. Earth in Upheaval. New York, NY: Doubleday, 1955 p. 121
[162] Fontenay, Charles. Estes Kefauver: A Biography. Olympic Marketing, 1980, p. 295
[163] Hapgood, Charles. The Path of the Pole. Philadelphia, PA: Chilton Books, 1970 p. 43
[164] Hapgood, The Path of the Pole. p. xi
[165] Hapgood, The Path of the Pole, p. 205
[166] Lemieux, Amy. "Everything You Didn't Want To Know About The Polar Shift." The Republican Standard, February 1, 2018
[167] Jardetsky, Wenceslas. "Aperiodic pole shift and deformation of the Earth's crust." Journal of Geophysical Research, vol. 67, issue 11, October 1962
[168] Cao, Q. & Wang, P. "Stokes-Einstein relation in liquid iron-nickel alloy up to 300 GPa." Journal of Geophysical Research, vol. 122, issue 5, May 2017
[169] White, John. Pole Shift: Scientific Predictions and Prophecies About the Ultimate Natural Disaster. Virginia Beach, VA: A.R.E. Press, 1980, p. 95
[170] Krieger, Douglas. Signs in the Heavens and on the Earth. CreateSpace, 2014
[171] Thomas, Chan. The Adam and Eve Story. Los Angeles, CA: Emerson House, 1965 p. 26
[172] https://www.cia.gov/library/readingroom/docs/CIA-RDP79B00752A000300070001-8.pdf
[173] Montaigne, David. End Times and 2019. Kempton, IL: Adventures Unlimited Press,

2013 pp. 328-334
[174] Michell, John. The Dimensions of Paradise: Sacred Geometry, Ancient Science, and the Heavenly Order on Earth. Rochester, VT: Inner Traditions, 2008 p. 50
[175] Lemesurier, Peter. The Great Pyramid Decoded. Rockport, MA: Element Books, 1996, p. 146
[176] Weidner, Jay & Vincent Bridges. The Mysteries of the Great Cross of Hendaye: Alchemy and the End of Time. Rochester, VT: Inner Traditions, 2003 p. 378
[177] "About geomagnetic reversal and poleshift." 3/15/2011, https://watchers.news/2011/03/15/about-geomagnetic-reversal-and-poleshift/
[178] Martin, Sean. "Does melting snow reveal ancient human settlement in Antarctica?" The Express. London, UK: 3.9/2018 https://www.express.co.uk/news/weird/929387/ancient-human-settlement-Antarctica-piri-reis-map
[179] Montaigne, David. "Third Secret of Fatima – Prophecy of a Pole Shift" 6/21/2016 http://beforeitsnews.com/prophecy/2014/06/third-secret-of-fatima-prophecy-of-a-pole-shift-2462116.html
[180] De la Sainte Trinite, Michel. The Whole Truth About Fatima: The Third Secret. Volume III. Buffalo, New York 1990. pp. 578–579
[181] http://www.bibliotecapleyades.net/profecias/esp_profecia05.htm
[182] originally from the October 1981 issue of Stimme des Glaubens
[183] http://www.tldm.org/News10/MalachiMartinBelievedInBayside.htm
[184] http://garabandalnews.overblog.com/2014/01/fatima-3rd-secret-and-facts.html
[185] www.teslasociety.com/tunguska.htm
[186] Wood, Dr. Robert M. "McDonnell Douglas studied UFOs in 1960s: Project called BITBR for 'Boys in the Back Room'" MUFON UFO Journal, Vol. 486, October 2008, p. 4 - http://documents.theblackvault.com/documents/MUFON/Journals/2008/October_2008.pdf
[187] Thomas, Chan. The Adam and Eve Story. Los Angeles, CA: Emerson House, 1965 p. 6
[188] https://www.cia.gov/library/readingroom/docs/CIA-RDP79B00752A000300070001-8.pdf
[189] Same site at cia.gov
[190] White, John. Pole Shift: Scientific Predictions and Prophecies About the Ultimate Natural Disaster. Virginia Beach, VA: A.R.E. Press, 1980 p. 273
[191] Creighton, Scott and Gary Osborn. The Giza Prophecy: The Orion Code and the Secret Teachings of the Pyramids. Rochester, VT: Bear & Company, 2012 p. 263
[192] Thomas, Chan. The Adam and Eve Story. Los Angeles, CA: Emerson House, 1965 p. 12
[193] Thomas, The Adam and Eve Story. p. 13
[194] https://www.cia.gov/library/readingroom/docs/CIA-RDP79B00752A000300070001-8.pdf
[195] Same site at cia.gov
[196] Thomas, Chan. The Adam and Eve Story. Los Angeles, CA: Emerson House, 1965 p.

[197] White, John. <u>Pole Shift: Scientific Predictions and Prophecies About the Ultimate Natural Disaster.</u> Virginia Beach, VA: A.R.E. Press, 1980 pp. 105-106

[198] LaViolette, Paul. <u>Earth Under Fire: Humanity's Survival of the Ice Age.</u> Rochester, VT: Bear & Company, 1997, p. 108

[199] Kroll, Henry. <u>Cosmological Ice Ages.</u> Bloomington, IN: Trafford, 2009 p. 359

[200] LaViolette, Paul. <u>Earth Under Fire: Humanity's Survival of the Ice Age.</u> Rochester, VT: Bear & Company, 1997 pp. 114-115

[201] LaViolette, <u>Earth Under Fire,</u> pp. 164, 194

[202] Flem-Ath, Rand & Rose Flem-Ath. <u>When the Sky Fell: In Search of Atlantis.</u> New York, NY: St. Martin's Press, 1995, pp. 20-21

[203] LaViolette, Paul. <u>Earth Under Fire: Humanity's Survival of the Ice Age.</u> Rochester, VT: Bear & Company, 1997, p. 161

[204] Weidner, Jay & Vincent Bridges. <u>The Mysteries of the Great Cross of Hendaye: Alchemy and the End of Time.</u> Rochester, VT: Inner Traditions, 2003, p.357

[205] Weidner, <u>The Mysteries of the Great Cross of Hendaye,</u> p. 357

[206] LaViolette, Paul. <u>Earth Under Fire: Humanity's Survival of the Ice Age.</u> Rochester, VT: Bear & Company, 1997, p. 72

[207] LaViolette, <u>Earth Under Fire,</u> p. 368

[208] LaViolette, <u>Earth Under Fire,</u> pp. 88-89

[209] Wilcock, David. antantipedia.ie/samples/tag/david-wilcock/

[210] Flem-Ath, Rand & Rose Flem-Ath. <u>When the Sky Fell: In Search of Atlantis.</u> New York, NY: St. Martin's Press, 1995, p. 7

[211] Creighton, Scott and Gary Osborn. <u>The Giza Prophecy: The Orion Code and the Secret Teachings of the Pyramids.</u> Rochester, VT: Bear & Company, 2012 p. 153

[212] "Unravelling the Mummy Mystery – Using DNA" Egyptology Online, April 10, 2009

[213] Hancock, Graham & Robert Bauval. <u>The Message of the Sphinx,</u> New York, NY: Three Rivers Press, 1996, p. 10

[214] Clow, Barbara Hand. <u>Catastrophobia.</u> Rochester, VT: Bear & Company, 2001, p. 191

[215] Creighton, Scott. "Pole Shift and the Pyramids: Are Strange Alignments on Egypt's Giza Plateau Clues to Ancient Catastrophe?" <u>Atlantis Rising Magazine,</u> #127, Jan/Feb 2018

[216] Dash, Glen. "The Great Pyramid Diagonals." 12/3/2014 http://glendash.com/blog/2014/12/03/the-great-pyramid-diagonals-do-they-point-to-a-hidden-inner-platform-within-the-pyramid/

[217] Flem-Ath, Rand & Rose Flem-Ath. <u>When the Sky Fell: In Search of Atlantis.</u> New York, NY: St. Martin's Press, 1995, p. 24

[218] Holsinger, Rosemary. <u>Yurok Tales.</u> Etna, CA: Bell Books, 1992 p. 68

[219] LaViolette, Paul. <u>Earth Under Fire: Humanity's Survival of the Ice Age.</u> Rochester, VT: Bear & Company, 1997 p. 334

[220] https://www.youtube.com/watch?v=4RH9CFRr1RY

[221] Michell, John. <u>The Dimensions of Paradise: Sacred Geometry, Ancient Science, and</u>

the Heavenly Order on Earth. Rochester, VT: Inner Traditions, 2008, p. 11
[222] De Santillana, Giorgio & Hertha von Dechend. Hamlet's Mill: An Essay Investigating the Origins of Human Knowledge and its Transmission Through Myth. Jaffrey, NH: David R. Godine, 1969 p. 82
[223] Frawley, David. Astrology of the Seers: A Guide to Vedic/Hindu Astrology. Detroit, MI: Lotus Press, 1990 p. 48
[224] Rig Veda 8:6.39, 8:65.2f, and 9:7.6
[225] De Santillana, Giorgio & Hertha von Dechend. Hamlet's Mill: An Essay Investigating the Origins of Human Knowledge and its Transmission Through Myth. Jaffrey, NH: David R. Godine, 1969 pp. 140-141
[226] LaViolette, Paul. Earth Under Fire: Humanity's Survival of the Ice Age. Rochester, VT: Bear & Company, 1997 p. 322
[227] Weidner, Jay & Vincent Bridges. The Mysteries of the Great Cross of Hendaye: Alchemy and the End of Time. Rochester, VT: Inner Traditions, 2003, p. xiii and p. 1
[228] Weidner, The Mysteries of the Great Cross of Hendaye, p. 360
[229] Weidner, The Mysteries of the Great Cross of Hendaye. p. 337
[230] Weidner, The Mysteries of the Great Cross of Hendaye, pp. 4-6
[231] Weidner, The Mysteries of the Great Cross of Hendaye. p. 349
[232] Weidner, The Mysteries of the Great Cross of Hendaye. p. 350, containing quote from Graham Hancock's Fingerprints of the Gods, p. 495
[233] Weidner, The Mysteries of the Great Cross of Hendaye, p. 2
[234] http://www.poleshiftnews.com/how-serious-is-a-pole-shift.html
[235] Montaigne, David. "Enoch and Elijah – Witnesses to POLE SHIFTS" 10/16/2014 http://beforeitsnews.com/prophecy/2014/10/enoch-and-elijah-witnesses-to-pole-shifts-2464786.html
[236] Bruchac, Joseph & Michael Caduto. Native American Stories. Golden, CO: Fulcrum Publishing, 1991, p. 112
[237] Bruchac, Native American Stories. p. 23
[238] Campbell, Joseph. The Mythic Image. Princeton, NJ: Princeton University Press, 1974, p. 29
[239] Jenkins, John M. Maya Cosmogenesis 2012. Santa Fe, NM: Bear & Company, 1998, p. 127
[240] Jenkins, Maya Cosmogenesis 2012, p. 290
[241] LaViolette, Paul. Earth Under Fire: Humanity's Survival of the Ice Age. Rochester, VT: Bear & Company, 1997, p. 354
[242] Houck, C.M. The Celestial Scriptures: Keys to the Suppressed Wisdom of the Ancients. iUniverse, 2002
[243] Montaigne, David. "Enoch Lived in Atlantis." 12/16/2014 http://beforeitsnews.com/prophecy/2014/12/enoch-lived-in-atlantis-2466054.html
[244]http://en.wikipedia.org/wiki/Day_length#mediaviewer/File:Hours_of_daylight_vs_latitude_vs_day_of_year_cmglee.svg
[245] http://www.timeanddate.com/sun/israel/jerusalem?month=6
[246] Cayce, Edgar. Reading 5748-6, 7/1/1932

247 Montaigne, David. Antichrist 2016-2019, CreateSpace, 2014, p. 138
248 Michell, John. The Dimensions of Paradise: Sacred Geometry, Ancient Science, and the Heavenly Order on Earth. Rochester, VT: Inner Traditions, 2008, pp. 14-15
249 Michell, John. The New View Over Atlantis. San Francisco, CA: Harper & Row, 1983, p. 157
250 Campbell, Joseph. The Mythic Image. Princeton, NJ: Princeton University Press, 1974, p. 149
251 https://www.bibliotecapleyades.net/profecias/esp_profecia01i2.htm
252 Massey, Gerald. The Historical Jesus and the Mythical Christ. Glasgow, UK: Hay Nisbet, 1888?, p. 108-109
253 Hancock, Graham & Robert Bauval. The Message of the Sphinx, New York, NY: Three Rivers Press, 1996, p. 241
254 De Santillana, Giorgio & Hertha von Dechend. Hamlet's Mill: An Essay Investigating the Origins of Human Knowledge and its Transmission Through Myth. Jaffrey, NH: David R. Godine, 1969 p. 57
255 http://educate-yourself.org/cn/planetXsequel15jul;04.shtml
256 Montaigne, David. "Nibiru: It's Not What Most People Think It Is." 1/27/2015 http://beforeitsnews.com/prophecy/2015/01/nibiru-its-not-what-most-people-think-it-is-2466872.html
257 LaViolette, Paul. Earth Under Fire: Humanity's Survival of the Ice Age. Rochester, VT: Bear & Company, 1997, pp. 51 and 67
258 LaViolette, Earth Under Fire, p. 292
259 LaViolette, Earth Under Fire, p. 73
260 Michell, John. The Dimensions of Paradise: Sacred Geometry, Ancient Science, and the Heavenly Order on Earth. Rochester, VT: Inner Traditions, 2008, pp. 114-115
261 Michell, John. The Dimensions of Paradise: Sacred Geometry, Ancient Science, and the Heavenly Order on Earth. Rochester, VT: Inner Traditions, 2008, p. 22
262 Michell, The Dimensions of Paradise, p. 108
263 LaViolette, Paul. Earth Under Fire: Humanity's Survival of the Ice Age. Rochester, VT: Bear & Company, 1997,, p. 3
264 Jenkins, John M. Maya Cosmogenesis 2012. Santa Fe, NM: Bear & Company, 1998, p. 130
265 Gray, John. Near Eastern Mythology. London, UK: Hamlyn House, 1969. pp. 31, 32
266 Astrolabe B, the Star catalogue known as "KAV 218B ii, lines 29-32
267 https://www.youtube.com/watch?x-yt-ts=1422579428&v=rQiQfdde_sI&x-yt-cl=85114401#t-784
268 http://www.skepdic.com/sitchin.html
269 Jenkins, John M. Maya Cosmogenesis 2012. Santa Fe, NM: Bear & Company, 1998, p. 177
270 http://www.sitchiniswrong.com/nibiru.pdf
271 De Santillana, Giorgio & Hertha von Dechend. Hamlet's Mill: An Essay Investigating the Origins of Human Knowledge and its Transmission Through Myth. Jaffrey, NH: David R. Godine, 1969, p. 325

[272] Freedman, Immanuel. "The Marduk Star Nebiru," Cuneiform Digital Library Bulletin 2015:3, November 8, 2015
[273] Schlegel, Gustave. Uranographie Chinoise, 1875, p. 740
[274] Huang, Junjie & Erik Zurcher, Time and Space in Chinese Culture, New York, NY: E.J. Brill, 1995, p. 138
[275] Bodde, D. "Harmony and Conflict in Chinese Philosophy" in Studies in Chinese Thought, edited by A.F. Wright, Chicago, IL: University of Chicago Press, 1953, p. 26
[276] Brown, Hugh A. Cataclysms of the Earth. New York, NY: Twayne Publishers, 1967,, p. 15
[277] Montaigne, David. End Times and 2019. Kempton, IL: Adventures Unlimited Press, 2013, p. 183 – original web site not found
[278] De Santillana, Giorgio & Hertha von Dechend. Hamlet's Mill: An Essay Investigating the Origins of Human Knowledge and its Transmission Through Myth. Jaffrey, NH: David R. Godine, 1969 , pp. 158-159
[279] De Santillana, Hamlet's Mill, pp. 160-161
[280] De Santillana, Hamlet's Mill, p. 236
[281] Hancock, Graham & Robert Bauval. The Message of the Sphinx. New York, NY: Three Rivers Press, 1996, p. 201
[282] Flem-Ath, Rand & Rose Flem-Ath. When the Sky Fell: In Search of Atlantis. New York, NY: St. Martin's Press, 1995, pp. 55-56, citing Bibby, Geoffrey. Looking for Dilmun. New York, NY: Knopf, 1969 page not cited
[283] Flem-Ath, When the Sky Fell, p. 66, citing Tilak, The Arctic Home in the Vedas, p. 72
[284] Donnelly, Ignatius. Ragnarok. 1883 http://www.sacred-texts.com/atl/rag/index.htm
[285] Flem-Ath, Rand & Rose Flem-Ath. When the Sky Fell: In Search of Atlantis. New York, NY: St. Martin's Press, 1995, p. 65
[286] Lemesurier, Peter. The Great Pyramid Decoded. Rockport, MA: Element Books, 1996, p. 314
[287] Lemesurier, The Great Pyramid Decoded, p. 313
[288] Bruchac, Joseph & Michael Caduto. Native American Stories. Golden, CO: Fulcrum Publishing, 1991, p. 71
[289] Flem-Ath, Rand & Rose Flem-Ath. When the Sky Fell: In Search of Atlantis. New York, NY: St. Martin's Press, 1995, p. 26 – citing Bancroft's The Native Races, vol. 3, p. 154
[290] Flem-Ath, When the Sky Fell, pp. 28-30
[291] Flem-Ath, When the Sky Fell, p. 66, citing Tilak, The Arctic Home in the Vedas, p. 419
[292] Montaigne, David. End Times and 2019. Kempton, IL: Adventures Unlimited Press, 2013, p. 189
[293] Hays, H.R. In the Beginnings. New York, NY: G.P. Putnam's Sons, 1963 – no page given in citation in Mary Ellen Carter's Edgar Cayce on Prophecy, New York, NY: A.R.E. Press, 1968, p. 124
[294] Gilbert, Adrian & Maurice Cotterell. The Mayan Prophecies. New York, NY: Barnes

& Noble Books, 1995, p. 62

[295] Jenkins, John M. Maya Cosmogenesis 2012. Santa Fe, NM: Bear & Company, 1998, p. 323

[296] Campbell, Joseph. The Mythic Image. Princeton, NJ: Princeton University Press, 1974, p. 158

[297] Jenkins, John M. Maya Cosmogenesis 2012. Santa Fe, NM: Bear & Company, 1998, pp. 57, 59, 76

[298] Jenkins, Maya Cosmogenesis 2012, p. 28

[299] Achyra S., Suns of God: Krishna, Buddha, and Christ Unveiled. Kempton, IL: Adventures Unlimited Press, 2004, p. 86, citing Sir Arthur Weigall's The Paganism in Our Christianity, no page given

[300] Achyra S., Suns of God, p. 89

[301] Hall, Manly. The Secret Teachings of All Ages - An Encyclopedia Outline of Masonic, Hermetic, Qabbalistic and Rosicrucian Symbolic Philosophy. San Francisco, CA: H.S. Crocker, 1928, p. 119

[302] Godwin, Joscelyn. Atlantis and the Cycles of Time. Rochester, VT: Inner Traditions, 2011, p. 345 - paraphrasing Charles Dupuis' 1794 book: The Origin of All Cults

[303] De Santillana, Giorgio & Hertha von Dechend. Hamlet's Mill: An Essay Investigating the Origins of Human Knowledge and its Transmission Through Myth. Jaffrey, NH: David R. Godine, 1969 p. 57

[304] Santillana, Hamlet's Mill, pp. 87 & 146

[305] Santillana, Hamlet's Mill, p. 87

[306] Godwin, Joscelyn. Atlantis and the Cycles of Time. Rochester, VT: Inner Traditions, 2011, p. 351 - paraphrasing Jean Phaure's Le Cycle de L'Humanitie Adamique

[307] De Santillana, Giorgio & Hertha von Dechend. Hamlet's Mill: An Essay Investigating the Origins of Human Knowledge and its Transmission Through Myth. Jaffrey, NH: David R. Godine, 1969 p. 111

[308] Santillana, Hamlet's Mill, pp. 140-141

[309] Santillana, Hamlet's Mill, p. 82

[310] Hancock, Graham & Robert Bauval. The Message of the Sphinx. New York, NY: Three Rivers Press, 1996, pp. 166, 159

[311] LaViolette, Paul. Earth Under Fire: Humanity's Survival of the Ice Age. Rochester, VT: Bear & Company, pp. 80-81

[312] Wilson, Colin, & Rand Flem-Ath. The Atlantis Blueprint: Unlocking the Ancient Mysteries of a Long-Lost Civilization. New York, NY: Little, Brown & Company, 2000, p. 171

[313] LaViolette, Paul. Earth Under Fire: Humanity's Survival of the Ice Age. Rochester, VT: Bear & Company, p. 248

[314] De Santillana, Giorgio & Hertha von Dechend. Hamlet's Mill: An Essay Investigating the Origins of Human Knowledge and its Transmission Through Myth. Jaffrey, NH: David R. Godine, 1969, p. 398-399

[315] Clark, E.E. Indian Legends of Canada. Toronto: McClelland & Stewart, 1960, pp. 17-19

[316] Campbell, Joseph. The Mythic Image. Princeton, NJ: Princeton University Press, 1974, p. 202
[317] Jenkins, John M. Maya Cosmogenesis 2012. Santa Fe, NM: Bear & Company, 1998,, p. 135, 173
[318] Waters, Frank. Book of the Hopi. London, UK: Penguin Books, 1977, pp. 12-22
[319] Bruchac, Joseph & Michael Caduto. Native American Stories. Golden, CO: Fulcrum Publishing, 1991, pp. 65-66
[320] Bancroft, H.H. The Native Races, San Francisco, CA: The History Company, 1886, vol. 3, pp. 89-90
[321] Clark, E.E. Indian Legends of Canada. Toronto, Canada: McClelland & Stewart, 1960, pp. 20-21
[322] Alexander, H.B. North American Mythology, 1910, p. 117
[323] White, John. Pole Shift: Scientific Predictions and Prophecies About the Ultimate Natural Disaster. Virginia Beach, VA: A.R.E. Press, 1980, p. 278
[324] https://en.wikipedia.org/wiki/Blue_Star_Kachina also Waters, Frank. Book of the Hopi. London, UK: Penguin Books, 1977, pp. 21-22
[325] http://earthmysterynews.com/2016/05/28/the-hopi-blue-star-kachina-appears-again-the-final-sign-before-the-day-of-purification-photos/
[326] Michell, John. The Dimensions of Paradise: Sacred Geometry, Ancient Science, and the Heavenly Order on Earth. Rochester, VT: Inner Traditions, 2008, p. 19
[327] Bruchac, Joseph & Michael Caduto. Native American Stories. Golden, CO: Fulcrum Publishing, 1991, p. 18
[328] White, John. Pole Shift: Scientific Predictions and Prophecies About the Ultimate Natural Disaster. Virginia Beach, VA: A.R.E. Press, 1980, p. 235

Made in United States
Cleveland, OH
03 January 2025